Disasters and the Media

PETER LANG
New York • Washington, D.C./Baltimore • Bern
Frankfurt • Berlin • Brussels • Vienna • Oxford

Library of Congress Cataloging-in-Publication Data

Pantti, Mervi.
Disasters and the media / Mervi Pantti,
Karin Wahl-Jorgensen, Simon Cottle.
p. cm. — (Global crises and the media; v. 7)
Includes bibliographical references.
1. Disasters—Press coverage. 2. Reporters and reporting.
3. Television broadcasting of news. 4. Journalism—Objectivity.
I. Wahl-Jorgensen, Karin. II. Cottle, Simon. III. Title.
PN4784.D57P36 070.4'3—dc23 2012006084
ISBN 978-1-4331-0826-6 (hardcover)
ISBN 978-1-4331-0825-9 (paperback)
ISBN 978-1-4539-0820-4 (e-book)
ISSN 1947-2587

Bibliographic information published by **Die Deutsche Nationalbibliothek.**
Die Deutsche Nationalbibliothek lists this publication in the "Deutsche Nationalbibliografie"; detailed bibliographic data is available on the Internet at http://dnb.d-nb.de/.

Cover photo used by permission of Lehtikuva / AFP / Jiji Press

The paper in this book meets the guidelines for permanence and durability of the Committee on Production Guidelines for Book Longevity of the Council of Library Resources.

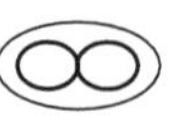

Printed in the United States of America

Disasters and the Media

Withdrawn from stock

Simon Cottle
General Editor

Vol. 7

This book is part of the Peter Lang Media and Communication list.
Every volume is peer reviewed and meets
the highest quality standards for content and production.

PETER LANG
New York • Washington, D.C./Baltimore • Bern
Frankfurt • Berlin • Brussels • Vienna • Oxford

Contents

SERIES EDITOR'S PREFACE:
Global Crises and the Media

We live in a global age. We inhabit a world that has become radically interconnected and interdependent and that communicates according to the formations and flows of the media. This same world also spawns proliferating, often interpenetrating, "global crises."

From climate change to the War on Terror, financial meltdowns to forced migrations; pandemics to world poverty, and humanitarian disasters to the denial of human rights, crises represent the dark side of our globalized planet. Their origins and outcomes are not confined behind national borders, and they are not best conceived through national prisms of understanding. The impacts of global crises often register across "sovereign" national territories, surrounding regions, and beyond. They can also become subject to systems of governance and forms of civil society response that are no less encompassing or transnational in scope. In today's interdependent world, global crises cannot simply be regarded as exceptional or aberrant events, erupting without rhyme or reason or dislocated from the contemporary world (dis)order. They are endemic to the contemporary global world, deeply enmeshed within it. And so, too, are they highly dependent on the world's media.

The series Global Crises and the Media sets out to examine not only the media's role in the *communication* of global threats and crises but also how the media can enter into their *constitution,* enacting them on the public stage and thereby helping to shape their future trajectory around the world. More specifically, the volumes in this series seek to: (1) contextualize the study of global crisis reporting in relation to wider debates about the changing flows and formations of world media communication; (2) address how global crises become variously communicated and contested in the media around the world; (3) consider the possible impacts of global crisis reporting on public awareness, political action, and policy responses; (4) showcase the very latest research findings and discussion from leading authorities in their respective fields of inquiry; and (5) contribute to the development of theory and debate that deliberately move beyond national parochialism and/or geographically disaggregated research agendas. In these ways the specially commissioned books in the Global Crises and the Media series aim to provide a sophisticated and empirically engaged understanding of the media's role in global crises and thereby contribute to academic and public debate about some of the most significant global threats, conflicts, and contentions in the world today.

Disasters and the Media by Mervi Pantti, Karin Wahl-Jorgensen and Simon Cottle addresses all of the series themes. The mediation of disasters is not only a critical issue, they argue, but also a timely one. In addition to disasters having become more common and often more catastrophic, they are also more mobile, crossing geographical boundaries and reverberating around the world in terms of impacts and emotional, humanitarian and political responses. In an increasingly global age, an age mediated in and through local-global communication flows, overlapping news formations, and proliferating social media, disasters now "hit home" for increasing numbers of us around the world almost instantaneously, in near real time. The different roles performed by media and communications within the public elaboration of and engagement with disasters, as well as in relation to the surrounding play of social interests and political power, demand increased recognition and careful theorization.

How are different disasters constituted in today's complex media environment? How are we invited to witness them and respond by these means? And how do the news media make disasters culturally meaningful and politically important? These fundamental questions underpin this intervention into the academic field of disaster study and scholarship. In a context in which both the nature of disasters and the media are changing rapidly, a fuller understanding of media's diverse roles and responsibilities in the communication of disasters, argue the authors, has become critical. Drawing upon four different but interrelated angles of approach—*globalization*, *new communication technologies*, *citizenship* and *emotion*—Pantti, Wahl-Jorgensen and Cottle explore and theorize the roles of the media in mediating disasters. *Disasters and the Media* opens up a distinctive and more nuanced understanding of the multidimensional complexity and shaping dynamics of contemporary mediated disasters in national and global contexts and does so based on original research on and discussion of some of the most momentous disasters of recent times.

Simon Cottle, Series Editor

Acknowledgments

THE AUTHORS WOULD LIKE TO THANK COLLEAGUES AND FRIENDS WHO HAVE IN their different ways contributed to the production of this book.

Simon Cottle would like to thank the following correspondents and journalists for graciously giving of their time, knowledge and experience of all things disaster reporting: Brendan Bannon, photojournalist; Jane Deith, Channel 4 reporter; Matthew Eltringham, BBC associate editor, interactivity and social media development; Deb Evans, BBC news editor; Andy Gallacher, BBC, North America correspondent; Chris Hogg, BBC, Shanghai correspondent; Robert Nisbet, Sky, U.S. correspondent; Sally Reardon, APTN, editor; and Janet Harris, independent documentary maker. Special thanks are also due to colleagues in the School of Journalism, Media and Cultural Studies at Cardiff University: Deborah Lloyd for expertly transcribing the interviews that formed the basis of Chapters 5 and 9; Emma Gilliam, Richard Tait, and Colin Larcombe, who all helped facilitate these interviews; and Justin Lewis, head of school, for supporting this project.

Karin Wahl-Jorgensen wishes to thank her many wonderful colleagues at Cardiff University who have helped shape her thinking on this project in various ways, including Paul Bowman, Stephen Cushion, Iñaki Garcia-Blanco, Emma Gilliam, Matt Hills, Justin Lewis, David Machin, Kerry Moore, Verica Rupar, Andy Williams, and Tamara Witschge, as well as those elsewhere, including Folker

Hanusch, Stijn Joye, Bella Mody, Mirca Madianou, Patrick Lee Plaisance and Barry Richards.

Mervi Pantti would like to thank all her great colleagues in the Department of Social Research, Media and Communication Studies, at the University of Helsinki, and special thanks are due to colleagues in the International Master's programme Media and Global Communication, Marko Ampuja and Yonca Ermutlu, and Hannu Nieminen, head of Media and Communication Studies, for their support and friendship. She would also like to acknowledge the Helsingin Sanomat Foundation for its financial support.

And finally all the authors would like to thank Mary Savigar, commissioning editor at Peter Lang, for her patience and welcome support as this project progressed and finally came to fruition.

1 Introduction

Disasters and the Media: Why Now?

WE LIVE IN AN AGE INCREASINGLY DEFINED BY MEDIA AND DISASTERS. IT IS THROUGH the media that most of us encounter disasters and bear witness to suffering around the world. In recent years we have witnessed a series of large-scale disasters, including the Indian Ocean tsunami in 2004, Hurricane Katrina in 2005, the cyclone in Burma and Sichuan earthquake in China in 2008, the devastating earthquakes in Haiti in 2010 and Japan in 2011, followed by a tsunami and nuclear meltdown at Fukushima. All have caused unforeseen losses of human life, massive social disruption and economic damage. According to Oxfam (2007), the total number of natural disasters over the last two decades has quadrupled, and during the same period more people have become affected by them. In industrialized societies, there is also clear evidence of the growth in the frequency of man-made disasters, such as fires, large-scale traffic accidents, industrial explosions, and major incidents of pollution. Moreover, in today's interconnected and interdependent world, a combination of powerful worldwide processes including climate change, population growth, urban migration and increased resource scarcity all exacerbate if not cause the likelihood of "disasters." There are strong grounds to expect that this trend toward greater vulnerability and human insecurity will only continue in the years ahead (Abbott, Rogers, & Sloboda, 2006; OECD, 2004; Oxfam, 2009a).

Current and future global crises are critically dependent on media and communications worldwide. Earlier research has shown that the role of media in disasters is both crucial and problematic. We respond to disasters through media images and discourses that invest them with emotional charge and wider importance. When mediated, disasters have the capacity to mobilize solidarities both within and beyond national borders. That media reports can position us as a global moral community and signal normative emotional responses is demonstrated in Secretary-General Ban Ki-moon's statement to the United Nations (UN) General Assembly following the Pakistan floods in August 2010 and broadcast around the world: "Pakistan is facing a slow-motion tsunami. Its destructive power will accumulate and grow with time.…Make no mistake: this is a global disaster, a global challenge. It is one of the greatest tests of global solidarity in our times" (United Nations, Secretary-General, 2010).

It should be noted, however, that disasters have created transnational media spaces and communities of feelings that overcome national boundaries before the development of "disaster journalism" in its current form in the 19th century and long before the modern audiovisual technologies that altered the scope and nature of media witnessing (Ellis, 2000). The 1755 Lisbon earthquake (followed by rampant fires and a tsunami), which killed 60,000 people, for example, sent shock waves across Europe. It has been characterized as "the first major news event" (Murteira, 2004), "the first media disaster" (Salmi, 1996), and "the first modern disaster" (Dynes, 2000, p. 113). The disaster remained in European newspaper headlines for several months and in public discussion for decades. Information, moral interpretations of the disaster, as well as graphic descriptions and dramatic accounts were first circulated in newspapers and public letters and later on in popular media texts such as broadside ballads, peep show spectacles, and theater shows across Europe (Murteira, 2004; Salmi, 1996).

The Lisbon disaster represented a turning point in mobilizing state-led emergency responses and demonstrations of transnational solidarity (Dynes, 2000; Murteira, 2004). Many European countries sent money, food, building supplies and labor to Portugal, and not unlike today's news media, newspapers were keen to report on the relief efforts. For instance, *The London Gazette* (March 10, 1756, cited in Murteira, 2004) reported that an aid ship was sent from the German city of Dantzick for the rebuilding of Lisbon: "Within these few Days past, a great Quantity of Timber, for building Houses, &c. has been shipped off from this Port to Lisbon," and *The Scots Magazine* (Vol. 18, March 1756, cited in Murteira, 2004) covered the arrival of British relief supplies and how they were going to be distributed among the people living in Lisbon: "Five of the Irish transports are arrived.

Dispositions are making for the distribution of the beef and butter among the poor, but this court has insisted upon the English subjects being first served." The international relief effort and extensive media interest were not only due to the enormous loss of life and devastation of the Lisbon quake but also to the commercial and political interests of other European countries in Portugal.

Parallels have been drawn between the Lisbon earthquake and the one which struck Japan in 2011, because of their scale and the complexity of their causes. While the Japanese disaster is in many ways an important example of the changing nature of global disasters, it also clearly accentuates the fact that disasters invariably become infused with diverse cultural meanings and political discourses that exceed the disaster itself. For instance, political parties and antinuclear advocates in Europe and elsewhere seized upon the Japanese disaster to raise the issue of nuclear safety, thus transforming the disaster into a political project at the national and global levels, as illustrated in the statement of Greenpeace EU nuclear policy adviser Jan Haverkamp: "It is important for Europeans to realize that you don't need a big earthquake to cause a nuclear catastrophe. It's time we moved away from dangerous and expensive nuclear and truly embraced renewable power" (cited in Cleaver, 2011).

For our purposes, it is also interesting to note that the Lisbon earthquake marked a watershed in the understanding of what a natural disaster is and what causes it. The traditional understanding was that disasters were signs of divine punishments for human sins (Kempe, 2003; Rohr, 2003). In the 18th century, the religious interpretation of natural disasters as signs of divine rage was challenged, and scientific explanations proposed that disasters should be seen as part of nature and thus "normal." The Lisbon earthquake became, in effect, a topic in intellectual debates over modernity, with profound implications for our own perception of disasters. Two of the main advocates of the Enlightenment, Rousseau and Voltaire, criticized the idea of divine punishment, but while Voltaire attributed the earthquake to nature, Rousseau propagated a view that disasters were ultimately a product of human agency, which created our vulnerability in the first place (Dynes, 2000). Anticipating current social science perspectives on disasters, Rousseau pointed to the idea that disasters are socially constructed and that their meaning depends on cultural context, for example, where they take place and who is affected by them: "You might have wished...that the quake had occurred in the middle of a wilderness rather than in Lisbon....But we do not speak of them, because they do not cause any harm to the Gentlemen of the cities, the only men of whom we take account" (cited in Dynes, 2000, p. 10) As Rousseau's comment suggests, our response to disaster depends not only on our nature or the extent of our suffering but is also linked to socially constructed notions of worthy and unworthy victims.

Today the role performed by media in the social construction of disasters proves no less consequential, though the nature of disasters and media and communications are now very different from the time of the "first media disaster" represented by the 1755 Lisbon earthquake.

The mediation of disasters is not only a problematic issue but also a timely one. In addition to disasters having become more common, more damaging, and potentially more catastrophic, they are also more mobile, crossing geographical boundaries and reverberating around the world in terms of impacts and humanitarian, emotional, and political responses. In an increasingly global age, an age mediated in and through local-global communication flows and overlapping news formations, disasters can "hit home" for increasing numbers of us around the world in near real time. The contemporary communications environment comprises a complex of traditional and mainstream mass media conveyed by new delivery systems and platforms and incorporating new contra-flows and cross-over productions from formerly peripheral nations. In addition, new communication technologies such as mobile telephony, the Internet and social media have now entered the daily communications mix and both interpenetrate with mainstream media and provide overlapping and alternative communication channels. When mediated in and through this rapidly evolving media and communication ecology, disaster communications promise to unsettle traditional hierarchies of communication power (Allan & Thorsen, 2009; Gowing, 2009; McNair, 2006).

New questions have arisen about the role of social media and citizen participation in circulating and constructing meanings of disasters and defining public knowledge, emotion and action, whether in global or national contexts. For example, the Asian tsunami in 2004 showed how "ordinary people" may have an increasingly important role to play in disaster coverage by providing immediate information and dramatic eyewitness images and accounts. The exponential increase in mobile telephony around the world in recent years played a key part in the relief efforts following the Haiti earthquake in 2010, alerting rescuers to the location of buried victims, communicating disaster relief services and raising money for relief efforts from donors around the world.

When mediated in the global marketplace of news, disasters today can rapidly become transacted locally, nationally and globally, with varied repercussions and impacts. We also know that while some disasters become subject to intense media exposure and invested with emotion and calls for help, many others around the world go relatively unreported. It seems that a terrible "calculus of death," based on crude body counts as well as proximities of geography, culture and economics, has become institutionalized in the professional judgments, practices and news values of the

Western media (Cottle, 2009a). Even so, media coverage can also help to open up public debate about preceding conditions of vulnerability and issues of disaster preparedness and responsibility.

This complexity and differentiation in the field of mediated disasters require enhanced recognition and theorization with respect to the different parts performed by media and communications within the constitution of disasters as well as in relation to the surrounding play of social interests and political power. Recent years have seen a growing sophistication of scholarly work on disasters. This includes work that theorizes disasters in the context of globalized risk society and which also looks at questions of global power relationships and the representational forms of disaster coverage. Disasters have variously been conceived, for example, as "cosmopolitan moments" (Beck, 2009) that encourage an awareness of global interdependency, "disaster shocks" (Klein, 2007) that are appropriated by political and economic elites to reassert forms of social and political control, and as "mediatized public crises" (Alexander & Jacobs, 1998) and "focusing events" (Tierney, Bevc, & Kuligowski, 2006) that give vent to contending discourses and projects for change in moments of heightened reflexivity. These and other theoretical conceptualizations prove productive, encouraging new vistas on the nature, forms and impacts of today's mediated disasters.

The starting point of this book, therefore, is that media today perform a leading role in the public constitution of disasters, conditioning how they become known, defined, responded to and politically aligned. When mediated, disasters are often narrated according to established cultural codes and scripts and invested with emotion. The "drama" and narratives of disaster, moreover, appear to unfold through a sequence of known stages or "acts," each populated by a cast of disaster "actors" who offer up expected "performances" on the "media stage." For example, victims and eyewitnesses provide emotional accounts and relive moments of terror and trauma; survivors are singled out for miraculously escaping from the jaws of death; rescuers can find themselves turned into heroes or even canonized as latter-day saints for embodying virtues of service and self-sacrifice; and elites and officials are obliged to go on disaster "walkabouts" in front of cameras, performing acts of public sympathy and demonstrating their command of the situation or, later, their public contrition for what has gone wrong. In the flows and forms of the media, then, disasters become defined and filled with demands, emotions and beliefs that resonate deep within the surrounding culture and "civil sphere" (Alexander, 2006a). In such ways, media and communications enter into their course and conduct, shaping their forms of public elaboration and engagement and channeling disaster responses—as humanitarian agencies dependent on public donations, political elites struggling to maintain control, and citizen activists and survivors struggling to be heard are keenly aware.

How are different disasters constituted in today's complex media environment? How are we invited to witness them and respond by these means? And how do the news media make disasters culturally mean and politically matter? These questions underpin this book. In a context where both the nature of disasters and media are fast changing, a fuller understanding of media's diverse roles and responsibilities in the communication of disasters has become critical. Our focus is not confined to the immediate interactions between political actors, source fields and the news media, important as these are, but also extends to the media's in-depth engagement of wider society through its inscriptions of cultural codes and normative outlooks. And this complexity includes differentiated forms of narratives, appeals and depictions, including the representation of those directly affected—disaster victims and survivors, those who seek to steer them—political elites and oppositional voices, as well as those who institutionally or individually are obliged to respond—nongovernmental organizations (NGOs) and publics. Our approach is indebted to preceding research on the relationship between media and disasters, and indeed we both acknowledge and incorporate findings and arguments from colleagues working in the contemporary research field. However, our deliberate choice of emphases throughout the book also structures our approach to mediated disasters and helps to distinguish it from others in the field.

First, we set out to conceptualize and situate disasters in the context of a rapidly globalizing and crisis-prone world and consequently problematize the deceptively simple idea of "disasters" as unforeseen and calamitous events. Not all disasters today can be related to preceding global conditions or causes but, increasingly, many can, and insofar as they become defined and reported in global news media, their repercussions and responses can outstrip their immediate geographical setting. How ideas of the national and the global variously become constituted through mediated disasters is a theme of growing importance in the context of debates about the "world risk society," "cosmopolitanism" (Beck, 2000, 2006) and "mediated suffering" (Chouliaraki, 2006; Moeller, 1999; Robertson, 2010), and we pursue these further.

Second, this book is compelled to take cognizance of and pursue the role of new communication technologies in the field of disaster communications. We examine how these are contributing to today's increasingly expansive and complex media ecology and, possibly, augmenting, unsettling or even reconfiguring relations of communication power in disasters and crises. Here we are interested in exploring how the changing communication environment variously enters into disaster communications, enhancing the range of views and responses that emanate from inside the disaster zone and how communication technologies, both old and new, are contributing to the transformation of disaster visibility around the world and with what possible repercussions on the relationship between audiences and "witnessing texts" (Frosh, 2006), which produce imagined worlds and impose demands on their addressees.

We are also interested in tracing, through engagement with scholarly approaches and a series of empirical case studies, how mediated disasters can offer crucial opportunities for negotiating questions of citizenship, governance and the mobilization and management of emotion. Specifically in the context of mediated disasters, this book takes a particular interest in emerging areas of scholarship which examine, as our third emphasis, the place of "ordinary people" in the newsworld, and, as our fourth, the role of emotions in the political and humanitarian realm. In the era of globalization, media coverage of disasters is seen as encouraging new forms of cosmopolitanism and, as such, raises key questions about the nature of citizenship and the role of emotion in shaping the public response to disasters.

Through disaster coverage, citizens are encouraged to feel and act in particular ways, which have profound significance not just for the concrete circumstances of disaster victims but also for understandings of appropriate disaster citizenship and the local, national and global power relations that it constructs. Further, disasters open up new spaces and opportunities for the contestation of hegemonic power formations, by giving voice to affected citizens and their expressions of anger and calls for accountability. At the same time and relatedly, questions of emotions come to the forefront in debates over media coverage of disaster: When approached in mediated and global contexts, questions of mediated compassion, the potential emergence of global citizenship and a global public sphere(s) and even, possibly, the stirrings of what Ulrich Beck (2006) discerns as a new "cosmopolitan outlook" (forced into being by new global threats) all come to the fore. More concretely, the book explores the emotional labor of journalists, in narrating the emotions generated by disasters and shaping those of audiences and also, at times, expressing their own. Shifting regimes of emotional discourse, in turn, challenge and reshape journalistic professional ideals, including the "strategic ritual" of objectivity (Tuchman, 1972), and call into question the structuring binary opposition of emotion and rationality.

Drawing upon these four different but interrelated angles of approach—*globalization*, *new communication technologies*, *citizenship* and *emotion*—we aim to add to the discussion regarding the role of the media today—and tomorrow—in mediating disasters. As our opening comments have suggested, media and communications have become profoundly implicated within contemporary disasters, both defining them and helping to discharge their effects worldwide. And, as we have intimated, disasters of all scales, from the local to the global, are also inescapably conditioned by and can impact on surrounding social structures, political interests and cultural meanings—and here too media and communications, both old and new, are no less deeply involved. We are not claiming to offer complete answers to the difficult and important questions we approach in this book. Our aim is rather to raise and explore these questions to open up a more nuanced understanding of the multidimensional complexity and shaping dynamics of contemporary mediated disasters in national and global contexts. As such, we hope that this work will lead to continued research and thinking in this area. The complexity of media responses as well as the profound implications that they may hold for a globalizing world of disasters, crises and catastrophes, we suggest, has only just begun to be recognized and theorized.

***

The book is divided into two sections. The chapters in the first section, "Perspectives on Disasters: Globalization, Citizenship and Emotion," lay down the broad conceptual and theoretical foundations that inform our understanding of and approach to disasters and the news media in a global media age. The second section, "Making Disasters Mean and Politically Matter," developing on this foundation, focuses on a number of important themes, exploring these in greater depth and with increased traction through empirical engagement and case studies.

Chapter 2, "Media and Disasters in a Global Age," opens the book's substantive themes by revisiting earlier ideas and approaches to disaster and communications and seeks to reconceptualize the latter in ways better suited to the changing global nature of our world and the reporting of global disasters and crises. It reviews key theoretical approaches to disaster and opens up some of the different roles and responsibilities enacted by the media within them. This includes discussion on how mediated disasters both *express* and sometimes also *enter into* the surrounding field of contending political interests and projects for change.

Chapter 3, "The Geopolitics of Disaster Coverage," follows up on some of these general theoretical claims by considering how the professional practices and cultural forms of contemporary journalism structure the representations of disaster coverage and how these representations can variously participate in constructing citizens as both members of the audience and the victims of disasters. It suggests, however, that the forms of coverage and, consequently, the particular forms of citizenship they enable, are still overwhelmingly grounded in the nation state and its place within global geopolitical power relations—notwithstanding the global possibilities of some disasters for cosmopolitan solidarity and the changing global communications environment.

Chapter 4, "Emotional Discourses in Disaster News," explores the relationship between the emotional and the political in disaster coverage, offering a further perspective on citizen involvement and professional practice from the vantage point of emotion. Drawing on recent research across disciplines, which challenges the contrasting of emotion against reason and explores the increasing prominence of emotions in public life, it suggests that emotional expression is central to the media witnessing and mobilization of moral action and therefore inseparable from politics. The chapter addresses questions of how emotions are embedded in disaster narratives and what political projects they potentially motivate, how emotional discourses in the news media shape local and global audiences' engagement with disasters, and how journalism's witnessing role is transforming in the increasingly affective public sphere.

The book's second section, "Making Disasters Mean and Politically Matter," opens with chapter 5, "Producing News, Witnessing Disasters." Based on a close reading of the accounts of journalists and correspondents who between them have covered most of the world's recent major disasters, the discussion explores a deep seated antinomy at the core of their professional practices and helps to explain why and how some disasters are recognized as important news stories, while others receive relatively little, if any, news interest. Ideas of witnessing are an established part of journalism's declared institutional aims and professional practices in the context of disasters; here we consider how this becomes both enacted and fractured through an engrained "calculus of death" and emergent "injunction to care."

Chapter 6, "Compassion, Nation and Cosmopolitan Imagination," focuses on the discourse of compassion and its role in motivating public response and forming national and global feeling communities. The chapter, taking the 2004 Asian tsunami as a case study, explores whether the national media can invite members of the audience to become cosmopolitan citizens who take responsibility for global tragedies.

Chapter 7, "Disaster Citizenship and the Assumption of State Responsibility," addresses the relationship between disasters and international political debate. It analyzes the coverage of the 2008 Burma cyclone and China earthquake, along with the 2010 Haiti earthquake, in terms of their political discourses on the role of the state on the one hand and the international community on the other. The chapter demonstrates that profoundly ideological metadiscourses about forms of governance and citizenship are central to disaster coverage and how the latter, therefore, is inherently politicized. In particular, what we refer to as the "assumption of state responsibility"—or that the nation state affected by the disaster is responsible for the care of its citizens—is central to these politicized narratives and shapes an ever-present discourse. Ultimately, a very particular form of cosmopolitan citizenship is constructed through and bounded by the nationally and geopolitically inflected narratives of disasters.

Chapter 8, "Anger and Accountability in Disaster News," analyzes expressions of anger in global and national disaster coverage, paying special attention to the role of anger in constituting citizenship and community in the wake of major disasters. Following emerging scholarship on public anger, the chapter argues that anger opens up a space for ordinary people to critique power holders, allowing victims and those affected by disasters to raise questions of systemic failure and blame.

Chapter 9, "Transformations in Disaster Visibility," the last in this section, explores today's fast-changing communication environment and how this is transforming the visibility of disasters around the world and potentially contributing to not only processes of representation and recognition but also responsibility. We address how social media have been embraced and incorporated into mainstream news production when reporting disasters and how the Web is influencing disaster images produced by photojournalists. With reference to Haiti and other recent disasters we consider the development of new communication hybrids based on old and new media and how these are communicatively enfranchising disaster survivors and giving rise to new technical and voluntary communities of tech-savvy helpers. And we also consider the role of satellite surveillance and remote-sensing technologies and how these are mapping and visualizing unfolding humanitarian disasters and sometimes, through their re-presentation within mainstream news media, powerfully entering into the political field of human rights and humanitarian disasters.

Finally, Chapter 10 brings the book to a close, briefly reviewing the principal themes, findings and debates that have informed the preceding discussion and underlining our core arguments.

PART I

PERSPECTIVES ON DISASTERS: GLOBALIZATION, CITIZENSHIP, AND EMOTION

2

Media and Disasters in a Global Age

THIS CHAPTER SETS OUT TO MAP IDEAS AND THEORETICAL APPROACHES THAT CAN help underpin critical engagement with disasters and crises in a global media age. We begin by revisiting established social science approaches to the deceptively simple question of "what is a disaster?" and challenging managerialist and technocratic understanding of disaster communications. Media and communications, we argue, increasingly *constitute* disasters, conditioning how they become known, responded to and politically aligned. This requires that we move beyond narrow communication research agendas focused on the restoration of disrupted "information flows" or public relations efforts to manage organizational reputations and public perceptions in terms of "effective crisis communications." Disasters, crises and catastrophes also can no longer be presumed as territorially bounded or discrete national events, seemingly erupting without warning to disrupt routines, norms and social order. When approached in global and media contexts, contemporary disasters, we argue, are often best conceived and theorized as *endemic* to, *enmeshed* within and potentially *encompassing* of today's globally interconnected (dis)order (Cottle, 2011a). This prompts a reconceptualization of what we generally understand by "disasters" in a global age and, again, calls for enhanced recognition and theorization of the complex parts performed by media and communications within their global constitution as well as within the surrounding play of social interests and political power.

To help get a better fix on how the political enters into and shapes mediated disasters, we next consider how different approaches to mediated disasters have sought to conceptualize and theorize the operations of political power and how these become transacted within and through them. Here we turn to a discussion of, respectively, "disaster shocks" (Klein, 2007), "focusing events" (Tierney et al., 2006) and "elite indexing" (Bennett, Lawrence, & Livingston, 2007), all ideas that help to secure a more politically engaged appreciation of how mediated disasters *express* but sometimes also *enter into* the surrounding force field of contending political interests and projects for change. However, under the onslaught of new communication technologies, including the powerful convergence of mobile telephony, the Internet and new social media, disaster communications are now rapidly expanding and growing more complex (McNair, 2006; Allan & Thorsen, 2009; Gowing, 2009) and, possibly, changing traditional relations of communication power within them. This, too, now needs to be factored into our appreciation of the politics of disasters.

Notwithstanding the continuing expansion of the Internet and rise of social media, we also need to recognize and theorize the ways in which mainstream media continue to play a leading and performative role in the public constitution of some disasters (incorporating new forms of disasters communications) and infuse disasters with cultural codes and appeals, feelings and beliefs, that resonate deep within the "civil sphere" (Alexander, 2006a). This sometimes opens up possibilities for political disruption and challenge, as well as the reinforcement of political dominance and consensus. Finally, therefore, the chapter considers theoretical and conceptual constructs of "disaster marathons" (Liebes, 1998), "mediatized public crises" (Alexander & Jacobs, 1998) and "cultural pragmatics" (Alexander, Giesen & Mast, 2006) that help open up to analysis something of the cultural complexities and diverse appeals embedded within mediatized disasters—themes that are developed further in "Part II, Making Disasters Mean and Politically Matter."

What Is a Disaster?

Commonsense ideas of a "disaster" as any event that has negative consequences quickly lose analytical traction when applied to such diverse phenomena as unexpected events in the natural environment (floods, fires, hurricanes, droughts, earthquakes, volcanic eruptions); technological and industrial failings (aviation crashes, train derailments, industrial accidents, toxic releases); politically precipitated crises and conflicts involving mass death, violence or attrition (wars, acts of terror, civil disobedience); and longer term and systemic failings (poverty, human rights abuses, environmental collapse). Entangled within the catchall term "disasters," therefore, are thorny issues of agency and intentionality, differences between latent and manifest disasters, between rapid onset events and slow-burn processes, and implicit judgments that have yet to be made about disaster thresholds and referents—whether in respect of scale of negative impacts, size of the social collectivities involved or the degree of system disruption caused (see Perry, 2007; Rodríguez et al., 2007). Most critically of all perhaps, reflection on the concept of disasters raises fundamental questions of claims-making and power, that is, of who defines what is a disaster, when and how and with what consequences—questions that are no less pertinent when applied to the specialist academic field of disaster study, as we shall hear.

The history of disaster research as an academic field of study and inquiry is relatively recent (Perry, 2007; Rodríguez et al., 2007; Tierney, 2007). Ronald Perry (2007) usefully distinguishes here between three areas of research that he terms the *classic approach,* the *hazards-disaster tradition,* and the *socially focused tradition.* Each in its own way seeks to impose analytical clarity and disciplinary boundary on "disasters" as an object of study in the social sciences. In the classic period of disaster research, principally in the aftermath of World War II and culminating with the establishment of the Disaster Research Group in 1952 under the auspices of the National Academy of Sciences in the United States, disasters were implicitly seen as events that "acted as a catalyst for what would now be described as a failure of the social system to deliver reasonable conditions of life" (Perry, 2007, p. 5). Whether focused in respect of systems breakdown prompted by the bombing of European and Japanese cities or the disruption to normal life caused by earthquakes or responses to airplane crashes, disasters in this approach generally became investigated in terms of societal and political responses and the reconstitution of social systems and norms following a "disruptive" event. Disasters in this tradition, therefore, were conceived "fundamentally as disruptions of routines" (Perry, 2007, p. 8; Stallings, 1998), and researchers generally emphasized how disasters could be characterized in terms of a cycle of stability-disruption-adjustment (Perry, 2007, p. 8).

The hazards-disaster tradition, in contrast, generally views each disaster as "an extreme event that arises when a hazard agent intersects with a social system" (Perry, 2007, p. 9) and paved the way for a more concerted exploration of preceding conditions of vulnerability—conceptualizing disasters as "the interface between an extreme physical event and a vulnerable human population" (Hewitt, 1988, cited in Perry, 2007, p. 9). In these terms, this tradition begins to move toward a more fully "social" understanding of disasters as "social phenomena," the last of Perry's three discerned traditions. Here, Enrico Quarantelli, for example, emphasizes multidimensional social aspects of disasters when he defined them as 1) sudden-onset occasions, that 2) seriously disrupt the routines of collective units, 3) cause the adoption of unplanned courses of action to adjust to the disruption, 4) have unexpected life histories designated in social space and time, and 5) pose danger to valued social objects (summarized in Perry, 2007, p. 10). Even the event of a disaster, as much as its physical location and time, is conceived as thoroughly "social" as signalled with the use of such terms as *occasions, disrupted routines, social space and time, valued social objects,* and so forth. Russell Dynes (1998) similarly emphasized the "social" when defining disasters as "occasions when norms fail" and that cause a community to engage in extraordinary efforts "to protect and benefit some social resource" (cited in Perry, 2007, p. 11).

Notwithstanding such efforts by disaster researchers to bring analytical precision and conceptual clarity to "disasters" as an object for social scientific inquiry, we can detect a reticence to engage more critically and theoretically with issues of power, structural determination and cultural performativity with respect to disaster communications. As Kathleen Tierney (2007) has observed, traditional approaches to disaster research have too long been defined by their applied and organizational focus and need to link to fields of environmental sociology and risk as well as focusing more critically on core sociological concerns of social inequality, diversity and social change. Established approaches to disasters conceived as "disruptive events" too easily suggest a normative acceptance of prevailing systems and norms rather than regarding them as structurally implicated in the reproduction of humanly injurious outcomes, routinized over the longer term and contributing to "permanent emergencies" or "unending disasters" that fall off the disaster researchers' radar. How we conceptualize "disasters," what's ruled in and what's ruled out, it seems, is not without political or ideological effects. Craig Calhoun (2008) makes a similar point when castigating the "Western cultural imaginary" encoded in news representations of "humanitarian emergencies" (often referred to as "humanitarian disasters"). Calhoun argued that the term *humanitarian emergencies*

> implies sudden, unpredictable events that require immediate action. But many "emergencies" develop over longer periods of time and are not merely predictable but are watched for weeks or months or years before they break into public consciousness or onto the agendas of policy makers. (p. 83)

This commonly accepted "emergency imagination," he suggests, is implicitly powered and ideological. It "reflects both the idea that it is possible and desirable to 'manage' global affairs and the idea that many if not all of the conflicts and crises that challenge global order are the result of exceptions to it" (Calhoun, 2008, p. 97). Not only does the fixation on disaster "events," then, tend to displace from view the normalized "abnormality" of profound inequality and systematically stunted life chances that constitute for many their ongoing disaster, it also becomes insufficiently attentive to those powered processes of claims-making by which some disasters, and not others, become publicly labelled as such and thereby positioned for various forms of intervention or response (Cottle, 2009a; Hawkins, 2008; Molotch & Lester, 1974; Stallings, 1995; Tierney, 2007).

Arjen Boin goes some way in meeting these objections when arguing for the inclusion of "disaster" under the more encompassing conceptualization of "crisis" (Boin, 2005; Boin & 't Hart, 2007). Boin proposed that disasters, in the contemporary era, are better conceived as a subclass of "crises" in that the latter "not only covers clear-cut disasters but also a wide variety of events, processes and time periods that may not meet the disaster definition" but which nonetheless "makes way for situations of threat and successful coping efforts" as well as "all processes of disruption that seem to require remedial action" (Boin, 2005, p. 161). Disasters in this sense, then, are crises that have gone bad. These ideas have recently been extended to international/global phenomena that Boin and his colleagues refer to as "trans-system social ruptures" (TSSRs; Quarantelli, Lagadec, & Boin, 2007).

TSSRs are said to be phenomena which a) jump across national, international and political boundaries, b) at speed, c) have no central or clear point of origin, d) are potentially catastrophic in terms of possible victims, e) cannot be resolved by local responses, and f) involve both formal organizations and informal networks (Quarantelli et al., 2007). This reads as a timely and promising conceptual development of "disasters" when situated in international and transnational contexts, but it also underplays the *constitutive* role of media and communications in today's disasters and under-theorizes the interdependencies between different TSSRs and their embedding within processes of globalization or what the social theorist Ulrich Beck referred to as "world risk society" (2000). Here latent risks and perceived threats—not only manifest disasters or TSSRs—profoundly condition the institutional and knowledge-based systems of late or "second modernity" (Beck, 2000, 2009).

We return to the relevance of Beck's social theoretical ideas in the context of global disasters reporting in a moment, but it is clear that much depends on how we define and delimit the very idea of disaster. When situated in global context, disasters largely evade earlier conceptual attempts to delimit them as objects of social scientific inquiry and conceived as unforeseen and disruptive events. The following elaborates on why this should be so, exploring further the idea of global disasters and crises within a globalized context before returning to the crucial role of media and communications in their global enactment.

Disasters, Crises, Catastrophes...in a Globalized World

The dramatic increase in the number of natural disasters in the last two decades reached a peak in 2010, with several disasters that were categorized under the UN's term, *great natural catastrophes:* the earthquakes in Haiti, Chile and China, the heat wave in Russia and the floods in Pakistan. At least some of these so-called "natural disasters" may be related to anthropogenic climate change (Global Humanitarian Forum, 2009) and can best be described, therefore, as "*un*natural disasters," but they can in any case be seen as unequally distributed and socialized hazards. So-called "natural disasters" have always disguised their socialized nature and unequal impacts around the globe in that, for example, buildings, not earthquakes, kill people, and risk reduction strategies invariably cost money. Hence natural "hazards only become disasters when they exceed a community's ability to cope" (Holmes & Niskala, 2007, p. 2; see also United Nations Environmental Program, 2001).

Not all disasters, whether natural or (*un*)natural, however, automatically find prominent news exposure and encourage donor funds for disaster relief operations. The vast majority of "uninsured lives" in the South, it seems, are not only cheap (Duffield, 2007) but also unnewsworthy. This is often explained in terms of impinging geo-political interests, national cultural outlooks, and the operation of foreign news values (Bacon & Nash, 2002; Benthall, 1993; Galtung & Ruge, 1965; Hawkins, 2008; International Federation of Red Cross and Red Crescent Societies, 2005; Moeller, 1999; Rotberg & Weiss, 1996; Seaton, 2005), as well as the journalistic "calculus of death" (Cottle, 2009a) ideas explored in depth later in this book.

If increasing numbers of disasters around the world, as suggested, are best approached under a broader conceptualization of crises, the latter needs to be situated in a global context. In other words, many of today's disasters and crises are spawned by today's condition of globality and its systemic production of endemic threats. These both express and intervene within processes of globalization, interdependency and the production of unequal life chances. Global crises and disasters are not simply, then, characterized as "global" or "transnational" in terms of their scale or reach, however, but also by their endemic and enmeshed global nature and required global forms of response. Zygmunt Bauman (2007), for example, referred to today's global "metaproblems" as deeply rooted in a "negatively globalised planet." He suggested the following: "On a negatively globalized planet all the most fundamental problems—the metaproblems conditioning the tackling of all other problems—are *global,* and being global they admit of no local solutions; there are not, and cannot be local solutions to globally originated and globally invigorated problems" (italics in the original, pp. 25–26). As a way of getting some traction on these ideas, it is useful at this point to briefly look at the idea of "complex emergencies." By doing so we can throw into sharper relief just how "globally originated" and "globally invigorated problems" now often outstrip the conceptualization of "complex emergencies" and its capacity to explain humanitarian disasters.

At first sight, the concept of "complex emergencies" appears to capture something of the complex interdependencies at work in situations of crisis, including humanitarian disasters. Developed in the post-Cold War period with its increased opportunity for humanitarian (and military) intervention in conflict-based humanitarian disasters, the notion of complex emergencies helped to reintroduce "the political" into the notion of "humanitarian emergency" (see Calhoun 2008). In this way, "conflict-generated emergencies" (Macrae & Zwi, 1994, cited in Keen, 2008, p. 1), or "humanitarian crises that are linked with large-scale violent conflict—civil war, ethnic cleansing and genocide," became distinguishable from natural disasters or "disasters caused primarily by drought, floods, earthquakes, hurricanes, tidal waves or some other force of nature" (Keen, 2008, p. 1). But, as we can begin to detect in such statements, the definition and conceptualization of "complex emergency" prove inattentive to how the "global" may also be at work in complex emergencies.

For example, large-scale conflicts are rarely self-contained but embroiled in changing global configurations of state power and exacerbated if not fuelled by interests and resources that invade from well beyond the conflict zone or even surrounding region (Dillon & Reid, 2000; Duffield, 2001, 2007; Kaldor, 2006; Shaw, 2005). In the post-Cold War period, "new wars" are the product of failed and failing states and involve complex webs of interests and identities that benefit from overseas trade connections and remittances from abroad. In this context, famine and environmental forces can be used to advance war aims and processes of "ethnic cleansing," and distinctions between interstate wars and intrastate conflicts, as well as between political conflicts and natural disasters, become less clear-cut. In these and other ways, global forces and interconnections can be rendered invisible through the more nationally or regionally based focus of "complex emergencies."

The increase in "natural disasters" in recent years further underlines the consequences of globalization and what Anthony Giddens referred to as globally "socialized nature" (Giddens, 1990) and what Ulrich Beck called global "manufactured uncertainty" (Beck, 1992), with anthropogenic climate change, alongside other globalizing forces, now contributing to new forms of "manufactured (in)security" (Beck, 2009). These include the exacerbating crises of water, food and energy shortages, forced migration, intensified tribal conflicts, state human rights violations as well as the global insecurity of transnational terrorism and new forms of Western "risk-transfer" warfare (Abbott et al., 2006; Amnesty International, 2009; Oxfam, 2009a, 2009b; Shaw, 2005). Many of today's disasters, in the context of globalization, therefore, now need to be conceptualized in global terms. Beck, more than any other social theorist today, posits global threats and crises centre stage in his theorization of the global age, extending ideas of "risk society" (1992) to "world risk society" (2000) and discerning at least three different axes of world crises that he terms, respectively, "*ecological, economic* and *terrorist* interdependency crises" (Beck, 2006, p. 22). Beck has not only helped to theorize the endemic nature of global crises as defining forces of the global age but also reflects on how they may yet unleash radical impulses in a new global "civilizational community of fate" (2006, p. 13):

> [T]he cosmopolitan outlook means that, in a world of global crises and dangers produced by civilization, the old differentiations between internal and external, national and international, us and them, lose their validity and a new cosmopolitan realism becomes essential to survival. (2006, p. 14)

This view of "enforced enlightenment" and "cosmopolitan realism" opens up the possibility that the manufactured uncertainties and manufactured insecurities produced by "world risk society" (Beck, 2000), prompt transnational reflexivity, global cooperation and coordinated responses (though these same processes may also prompt much else besides). These are themes that resonate closely with current debates and research on the transnational potential of disasters to open up opportunities for imagined community and international solidarity in response to mediated coverage of distant suffering (see Chouliaraki, 2006; Cottle, 2008; Ignatieff, 1998; Kyriakidou, 2008, 2009; Robertson, 2008, 2010; Stevenson, 2004).

To focus on the global nature of many of today's disasters and crises is not, therefore, to succumb to *catastrophism* or an unfounded pessimism that indulges a sense of the apocalyptic; it is to deliberately reconceptualize accidents, disasters and crises so as to take them seriously in a global context. Beck is quite clear, for example, that he has no desire to become the "Hieronymus Bosch of sociology" but rather sets out "to develop the existing theory and sociology of risk" through the *globalization perspective,* the *staging perspective,* and the *comparative perspective*—all focused in respect of the interdependency crises of environment, economy and terrorism (2009, pp. 19–20).

Similarly, Paul Virilio (2007), in his disquisition on the "integral accident," a term he coined on the basis of inverting Aristotle's original notion of the accident as something outside of the essential characteristics of a *substance,* argued that the "global accident" today is prefigured in the nature and thinking of contemporary "progress":

> Once upon a time the local accident was still precisely situated—as in the North Atlantic for the *Titanic.* But the global accident no longer is and its fallout now extends to whole continents, anticipating the *integral accident* that is in danger of becoming, tomorrow or the day after, our sole habitat, the havoc wreaked by Progress then extending not only to the whole geophysical space, but especially to timespans of several centuries. (p. 11).

Virilio, like Beck, then, is entitled to argue that, "Far from urging some 'millenarian catastrophism,' there is no question here of making a tragedy out of an accident with the aim of scaring the hordes as the mass media so often do, but only finally of taking accidents seriously" (2007, pp. 11–12). Though we may want to quarrel with Virilio's somewhat short-circuited view of the media's more complex relation to global crises and his demeaning view of media publics (Cottle, 2011a), we may agree that to recognize the endemic nature of contemporary global accidents, disasters and crises cannot be simplistically interpreted as a collapse into catastrophist thinking. And here the recent reflections of Slavoj Žižek (2009, 2010) are also no less pertinent. Reflecting on recent crises and catastrophes as well as the forthcoming disasters of energy, water and food shortages, Žižek anticipates that this is likely to produce a "new era of apartheid in which secluded parts of the world enjoying an abundance of food, water and energy are separated from a chaotic 'outside' characterized by widespread chaos, starvation and permanent war" (2009, p. 84). He caustically observes how the United States managed to find and spend $700 billion to stabilize the banking system in the financial disaster of 2008, but only $2.2 billion was actually delivered by the world's richest nations to the poorer nations to help develop their agriculture. Under conditions of food scarcity and world poverty, Giorgio Agamben's notion of *homo sacer* (of bare, precarious, life) now applies, he says, to vast swathes of the world's population. "If this sounds apocalyptic," argues Žižek, "one can only retort that we live in apocalyptic times" (2009, p. 92).

The accumulating evidence of proliferating "accidents," "disasters," "crises" and "catastrophes" around the world demands that many of them should be conceived and theorized in globally endemic terms, lending weight to Bauman's conceptualization of a "negatively globalised planet," Virilio's "global accidents," Beck's "global interdependency crises" and Žižek's anticipated "new era of apartheid"—notwithstanding their differing theorizations of the same. Evidence for such globally produced disasters and catastrophes is not difficult to find. It was documented, for example, in the International Panel on Climate Change (IPCC, 2007) reports, Kofi Annan's Global Humanitarian Forum's shocking calculations of 300 million people now being seriously affected by climate change each year, including 300,000 deaths and prediction of 500,000 deaths by 2030 (Global Humanitarian Forum, 2009); the precariousness of interlocking global financial systems, periodic meltdowns, and their devastating impacts on developing countries (United Nations, 2009); the alarming rise of weather-borne disasters (Oxfam, 2007) and vector-borne diseases (World Health Organization, 2007); and the latest world audit of human rights abuses and their interlinkage with these and other forms of world crises (Amnesty International, 2009). As Beck observed, anyone approaching "world risk society" today is "in danger of being overwhelmed by a flood of mostly undefined problems" (2009, p. 20).[1]

To add to this global complexity, global disasters and crises are not necessarily self-contained or discrete phenomena but often interpenetrate and/or mutate into related crises and disasters and exacerbate yet others. The mutability of real-world crises and disasters, therefore, is not always accurately captured by distinct categories of disasters (as we have already discussed in respect of earlier social science attempts to define "disasters," the notion of "complex emergency" or even "trans-system social ruptures"). For example, according to a series of recent United Nations reports, amongst the countless ramifications of the 2008 global financial and economic crisis are not only its deepening of world poverty leading to an increase of 200,000 to 400,000 child deaths (United Nations, 2009) but also its impact on the life chances of migrants and others through cutbacks to health care programs including HIV/AIDS services (United Nations Development Program, 2009), its impact on the increasing numbers of women entering the sex trade (United Nations Inter-Agency Project on Human Trafficking, 2009), and its endangering of the world's forests following a retraction in house building that led to shrinking investments in forest industries and forest management, contributing further to global warming (United Nations, Food and Agricultural Organization, 2009).

Climate change and the extreme weather events it causes now create a catalogue of disasters that also enfold into each other. Changing climate can exacerbate competition for land, water and food, creating conditions for civil strife and political instability and producing issues of national and international insecurity (International Institute of Strategic Studies, 2007) and human (in)security (Kaldor, 2007). Climate change can also lead to the incubation of deadly vector-borne diseases and produce forced migrations of environmental refugees. The IPCC has predicted that by 2080, climate change will lead to 1.1 to 3.2 billion people experiencing water scarcity, 200 to 600 million suffering hunger, and 2.7 million a year subject to coastal flooding, leading to 250 million permanently displaced people (International Panel on Climate Change, 2007). To take another example, a combination of global factors also contributed to the global food crisis of 2008. These included the following: increased demand for grain-intensive meat production in developing economies such as China and India, poor harvests exacerbated by climate change in others, the production of biofuels displacing food production in the South to support climate change policies in the North, and rising world energy costs. Together these contributed to a marked increase in world food prices and a global food crisis that impacted the most on the world's poor and led to food riots in scores of cities around the world (Oxfam, 2009a). The Japanese disaster of 2011, to take one last example, involved an unfolding complex of an interpenetrating earthquake, tsunami, nuclear meltdown, economic crisis and world oil price rise as well as a cloud of nuclear distrust that circumnavigated the globe and entered into national debates about energy policy in a context of climate change.

As all these examples begin to illustrate, it is important to keep the multidimensional, interpenetrating and mutating character of global crises and disasters in mind if we are to avoid dissimulating the complex global connections involved. We also need to recognize and theorize the critical centrality of media and communication in their public constitution and enactment, which is discussed next.

Theorizing Mediatized Disasters

> The question of what is a disaster actually is easily answered by those directly affected. Even in the academic discussion disasters are usually conceived of as clearly defined entities which have distinct causes and effects. Such a perspective, however, easily loses sight of the iterative quality of disastrous events as well as of the social constitutedness of disasters, which are recognized as such only through interpretive procedures. Moreover, as an inter-subjective social reality they are constituted through communicative practices alone. (Bergmann, Egner, & Wulf, 2009, p. 1)

Disasters, as Jörg Bergmann and his colleagues proclaim, are profoundly dependent upon communications and communicative practices. How they are signalled and symbolized, turned into spectacles or effectively rendered silent on the media stage, can have far-reaching consequences for the victims and survivors directly affected, for surrounding communities, and for the conduct of social relations and political power more widely. The latter, courtesy of media and communications systems, can now variously enter into or appropriate disasters, mobilizing interests and identities in and through them, and aligning them to various causes.

Again, Beck's recent emphasis on *staging* in world risk society provides a useful springboard for considering the role of media and communications in communicating risks and the crises borne by world interdependency. His ideas follow directly from his understanding of risks as the inherently contestable knowledge claims generated under conditions of manufactured uncertainty and manufactured insecurity.

> Risks are social constructions and definitions based upon corresponding relations of definition. Their existence takes the form of (scientific and alternative scientific) knowledge. As a result their "reality" can be dramatized or minimized, transformed or simply denied according to the norms which decide what is known and what is not. They are products of struggles and conflicts over definitions within the context of specific relations of definitional power, hence the (in varying degrees successful) results of stagings. (Beck, 2009, p. 30)

It is this core understanding of risk, then, that leads Beck to grant media staging with such central significance (1992, 2009), and it is this same theorization that posits the media with potential "political explosiveness" (Beck, 2009, p. 98). When staged in the media, Beck suggests, global disasters and crises can become "cosmopolitan events":

> Thus "cosmopolitan events" are highly mediatized, highly selective, highly variable, highly symbolic local and global, national and international, material and communicative reflexive experiences and blows of fate that transcend and efface all social boundaries and overturn the global order that holds sway in people's minds. (2009, pp. 70–71)

It is not necessary to accept Beck's central theoretical preoccupation with "risk" and consequent emphasis on the "anticipated catastrophes "of contemporary world risk society, however, to acknowledge how global crises and catastrophes materially unfolding in the world today have become critically dependent on processes of social construction (cultural mediation) in the news media. In what ways news staging of different global crises and disasters may provide the foundation for Beck's "cosmopolitan moments" and the "globalization of emotions" as he suggests, however, clearly demand detailed empirical inquiry and careful thought. The chapters that follow explore empirically and conceptually many of the multidimensional and overlapping complexities involved. Though Beck's writings are exceptional amongst theorists of globalization in forefronting global crises and media staging, we need to attend more closely to instances of disaster reporting, theorizing their different forms and attending to how news media variously position crises and disasters as events of public concern and possible political action.

We shall return to important and diverse issues of mediation in the representation of disasters and how these variously inhibit or possibly contribute to "cosmopolitan moments" and "globalization of emotions" throughout this book. But first it is also useful to introduce how questions of political power and the relations between news media and political authorities and political challengers (Wolfsfeld, 1997) have been theorized by media and communication theorists and how "the political" is thought to critically intervene within and shape the public elaboration of disasters. Here, then, we briefly review three distinct theoretical approaches to the political-media nexus and how each variously conceives political interests at work in shaping mediated disasters before considering how the nexus between institutionalized political elites and mass media is now being challenged by the arrival of new social media.

Disaster Shocks, Focusing Events, Elite Indexing—and the Civilian Surge

When staged in the world's news media, major disasters have variously been theorized as opportunities for elites to capitalize on the "disaster shock" of catastrophic events furthering corporate economic interests and established political goals (Klein, 2007), "focusing events" that condense wider cultural frames and discourses to soften up publics into accepting, for example, future militarized control of disastrous events (Tierney et al., 2006), and moments of "elite indexing" in which the media align their coverage to the prevailing political views and degree of consensus about what needs to be done (Bennett et al., 2007). Though each of these perspectives on disasters posits a strong (sometimes determining) relationship among political power, media and disasters, the mechanisms of its enactment and the degree of political conformity that the media are thought to exhibit in this relationship are theorized quite differently. Each nonetheless proves instructive for a more politically sensitized approach to mediated disasters.

In her book *The Shock Doctrine,* subtitled *The Rise of Disaster Capitalism,* Naomi Klein (2007) develops her thesis about the ways in which disasters and crisis can be put to work in the service of powerful corporate and government interests.

> That is how the shock doctrine works: the original disaster—the coup, the terrorist attack, the market meltdown, the war, the tsunami, the hurricane—puts the entire population into a state of collective shock. The falling bombs, the bursts of terror, the pounding winds serve to soften up whole societies much as the blaring music and blows in the torture cells soften up prisoners. Like the terrorized prisoner who gives up the names of comrades and renounces his faith, shocked societies often give up things they would otherwise fiercely protect. Jamar Perry and his fellow evacuees at the Baton Rouge Shelter (following Hurricane Katrina) were supposed to give up their housing projects and public schools.

> After the tsunami, the fishing people in Sri Lanka were supposed to give up their valuable beachfront land to hoteliers. Iraqis, if all had gone according to plan, were supposed to be so shocked and awed that they would give up control of their oil reserves, their state companies and their sovereignty to U.S. military bases and green zones. (p. 17)

Klein's thesis is arresting. It urges us to step back from the immediate effects of seemingly disparate crises and disasters to see the bigger picture of how they can become politically appropriated and put to work. Disasters shock societies into giving up that which in normal circumstances would be defended against the further encroachments of corporate capitalism and neoliberal governance. Here the nebulous notion of "disaster," discussed earlier, is nailed down not by specific types of destructive events or processes but rather by an overriding sense of the political interests that can both profit from and steer them. Klein's thesis usefully reminds us, then, of how disasters and collective traumas cannot be approached as if in a political vacuum. Politics and the political enter into them through every pore, both preceding and surrounding their destructive eruption into everyday life and then taking advantage of the trauma and confusion that they cause. But in a mediated age, we might reasonably argue, "disasters" affect more than those immediately caught up within their destruction, and they have to if wider reactions and responses are to become activated. In this respect, then, Klein's relative silence on the nature of media involvement in disasters is conspicuous. "While the disaster capitalism complex does not deliberately scheme to create the cataclysms on which it feeds (though Iraq may be a notable exception)," she only notes in passing how "there is plenty of evidence that its component industries work very hard indeed to make sure that current disastrous trends continue unchallenged" (2007, p. 427). And here the "creeping expansion of the disaster capitalism complex into media may prove to be a new kind of synergy," she suggests, "building on the vertical integration so popular in the nineties" and profiting from panic (2007, p. 427). This is, it has to be said, a generalizing thesis based on a synergistic view of economic and political dominance and corporate media as both benefiting from the same "disaster capitalism" logic. Disasters, in this political economy sense, are good for media ratings, readers and revenue.

Kathleen Tierney and her colleagues (2006) provide a more culturally nuanced and empirically focused discussion of "the political" in their study of how the news reporting of Hurricane Katrina perpetuated a number of "disaster myths" and "framed" the aftermath of the disaster in politically consequential and damaging ways—ways that can also be interpreted as supportive of U.S. military and government interests. The authors summarized their findings and principal argument as follows:

> Initial media coverage of Katrina's devastating impacts was quickly replaced by reporting that characterized disaster victims as opportunistic looters and violent criminals and that presented individual and group behavior following the Katrina disaster through the lens of civil unrest. Later, narratives shifted again and began to metaphorically represent the disaster-stricken city of New Orleans as a war zone and to draw parallels between the conditions in that city and urban insurgency in Iraq. These media frames helped guide and justify actions undertaken by military and law enforcement entities that were assigned responsibility of the postdisaster emergency response. The overall effect of media coverage was to further bolster arguments that only the military is capable of effective action during disasters. (pp. 60–61)

Based on this critical analysis of Katrina reporting, the authors argued that such media framing effectively serves to construct the disaster of Hurricane Katrina as a "focusing event" in which preceding and surrounding political discourses become condensed and communicated to the wider national society and thereby serve to legitimize the wider operations of political (and military) power. The study illuminates the damaging consequences of this media framing, reminding us that disaster survivors, contrary to popular myths, do not always succumb to panic or act with selfishness but can respond altruistically, helping neighbors and others. But these actions, tellingly, actually became thwarted by the imposition of military curfews and circulated media frames that discouraged outsiders from offering their help. This study also serves to remind us how new events, including disasters, can all too easily become inserted into established cultural frames and thereby emptied of their potential to assert new discourses and claims for recognition.

Both this study and Klein's thesis of "disaster shocks" advance relatively generalizing claims about the political appropriation of disasters by political authorities and their media framing in terms of political dominance. They provide us with a rather partial and possibly too static view of "the political" in disasters as constituted solely by systems of dominance and top-down structures of political, military and corporate power. In fact, as we shall see, disasters are also capable of sustaining different political outlooks and projects, some rooted in civil society and seeking opportunities for change. And this requires a more differentiated consideration of how disasters can become constructed and communicated differently in the media and in relation to the complex force field of social relations and political power seeking to influence media frames and agendas.

A model that begins to move in the direction of recognizing a more dynamic and politically contingent interface between news media and political and official elites is that of press-elite indexing (Bennett, 1990). This approach opens up for discussion the possibility that the news media can in fact entertain a more independent or even, on occasion, critical stance to the operations of political governance and power. According to the indexing model, the U.S. mainstream press normally report the news based on the sphere of official consensus and conflict, calibrating their stories according to the prevailing governmental and political field. Only exceptionally, when the political centre itself is divided and uncertain, do journalists feel capable of asserting a more independent and critical view (Hallin, 1986, 1994). In the case of Hurricane Katrina, Lance Bennett and his colleagues argued that the political vacation period that happened to coincide with the onslaught of Hurricane Katrina meant that officials were not able to manage the flow of information as effectively as they might normally have and that this created a rare "no spin zone" (Bennett et al., 2007, p. 64). The nexus between news media and government is theorized as a key for understanding how, in this particular case, a major national disaster became framed and opened up opportunities for public criticism and dissent. Interestingly, this finding, based on the output of different news outlets, appears to contradict somewhat those of Tierney et al. (2006). This in itself, of course, may tell us something about the differentiation that can be obtained within today's local-national media ecology, and both these studies remind us that reporting frames can in fact change through time and make different claims on readers and audiences in terms of how they become invited to respond to major disasters.

Each of the studies discussed earlier, and their informing theorizations of mediated disasters, clearly signals the operations of political power in their public shaping and how this generally privileges the interests of political authorities and dominant elites. This, as we have heard, can be conducted through the rapacious logic of neoliberal capitalism, exploiting disasters and media corporations seeking out their profitable synergies (Klein, 2007); the circulation of frames and cultural metaphors that already shape the political field and which serve to align disasters to dominant political projects and legitimize elite political control (Tierney et al., 2006); or the indexing of media to the prevailing views and consensus found in the political centre of society—and executed on the basis of routinized source dependencies and shared cultural values (Bennett et al., 2007).

In today's complex media ecology, however, we may want to inquire a little further into how globally expansive, interpenetrating, multimedia news flows unsettle, influence or simply circumvent traditional agenda setting, gatekeeping and elite indexing by national based news media (see Allan, 2006; Allan & Thorsen, 2009; Cottle, 2008; Cottle & Lester, 2011; Gowing, 2009; McNair, 2006; Seib, 2002; Volkmer, 2002). The arrival and rapid uptake (and customization) of new social media have also contributed to the "transformation of visibility" in contemporary societies (Thompson, 1995) and, according to Nik Gowing (2009), fundamentally shifted information power in crises. A "civilian surge" of information, he argues, is having an asymmetric, negative impact on the traditional structures of power. His ideas have particular relevance in the context of disaster communications and the possible reconfiguration of traditional relations of communication power:

> In a crisis there is a relentless and unforgiving trend toward an ever greater information transparency....Hundreds of millions of electronic eyes and ears are creating a capacity for scrutiny and new demands for accountability. It is way beyond the capacity and assumed power and influence of the traditional media. The global electronic reach catches institutions unaware and surprises with what it reveals. (Gowing, 2009, p. 1)

This "civilian surge," argues Gowing, unsettles the traditional monopoly on information and media by elites in times of crises and disasters. And in such pressurized moments, "the time lines of media action and institutional reaction are increasingly out of synch," precipitating potential public relations disasters with the fast release of unverified or insecurely sourced information into the public domain. Brian McNair's (2006) model of "cultural chaos" also theorizes the changing nature of news and power in a globalized world, and it does so on the basis of the perceived revolutionizing shift in communications complexity and its cultural dispersal of power. His thesis has particular relevance for the communication of crises and disasters seemingly qualifying critical presumptions about elite dominance and media control. According to McNair (2006), the capacity of elite groups to control the news agenda and define meanings is "effectively more limited than it has been since the emergence of the first news media in the sixteenth century" (p. 4).

As these and other influential voices now suggest, the changing communications environment is possibly altering relations of communication power, and indeed the nature of power itself, in a profoundly connected, mediated and networked world (Castells, 2008, 2009). In times of crisis and disaster, such transformations are likely to be all the more evident, thrown into sharp relief by the upsurge of on-the-spot, locally-globally dispersed, horizontal, bottom-up as well as vertical, top-down, public relations controlled communication flows (see Chapter 9). Of course these are challenging propositions, suggesting that the operation of elite power and its buttressing and enactments in and through the traditional elite-media nexus have become progressively eroded if not superseded in the new complexity and "cultural chaos" (McNair, 2006) of contemporary communications. Though many can see that the communications environment is now fast-changing, to what extent this is really altering underlying structures, ideologies and communication power in favour of a new "civilian surge," or in fact simply accommodates and incorporates the latter within continuing systems of elite-media dominance, remains a key point of contention. Politics and the political, then, now variously enter into the communication of disasters and are theorized and contested in no less politically engaged ways by scholars of media and disasters.

But political enactments of power do not only enter into the communication of disasters at the interface among media, political elites and the new civilian surge facilitated by emerging social media—however conceived and theorized. The political also enters disasters in and through the cultural work performed by the news media when making disasters mean. Here we need to attend more closely to how major disaster events are rendered culturally meaningful in the media and thereby politically consequential, which is discussed next.

Disaster Marathons and Mediatized Public Crises

> We sense a retreat from the genres of "media events" (Dayan & Katz, 1992)—the ceremonial Contests, Conquests and Coronations that punctuated television's first 50 years—and a corresponding rise in the live broadcasting of disruptive events such as Disaster, Terror and War. (Katz & Liebes, 2007, p. 157)

Elihu Katz, coauthor of the celebrated study *Media Events* (Dayan & Katz, 1992), and Tamar Liebes, author of "Television's Disaster Marathons" (1998), point to the rise of live disruptive events, including disasters, in contrast to ceremonial and integrative events, in recent television history. We will unpack the continuing significance of so-called "media events" for an appreciation of both the political and the cultural in mediated disasters presently. But first here is Liebes, reflecting on the cultural and political power of mediated disasters now conceived as disaster marathons:

> In contradistinction to media events, the shared collective space created by disaster time-out, zooming in on victims and their families, is the basis not for dignity and restraint but for the chaotic exploitation of the pain of participation on screen, and for the opportunistic fanning of establishment mismanagement, neglect, corruption, and so on. Whereas the principle of broadcast ceremony is to highlight emotions and solidarity and to bracket analysis, a disaster marathon constitutes a communal public forum where tragedy is the emotional motor which sizzles with conflict, emphasizing anxiety, argument and disagreement. (Liebes, 1998, pp. 75–76)

Liebes' treatise, based on a cultural reading of live disaster "events," offers a radical departure from the theoretical formulations discussed earlier. It suggests how the ongoing reporting of major disasters over time can in fact open up, rather than close down, contending political voices and issues to public view, with disasters acting as the vehicle for expressions of public dissent and disagreement, undermining authority claims for legitimacy and consensus. It also points to the possibility that in times of disaster, the media may be able to assert, on occasion, a more independent or even steering role in respect of the public orchestration of emotions and contentious issues.

The ideas of Jeffrey Alexander and Ronald Jacobs (1998) on "mediatized public crises," like those of Liebes, also take their departure from the seminal ideas of "media events" (Dayan & Katz, 1992) and further point to the political opportunities opened up by such mediatized crises:

> Celebratory media events of the type discussed by Dayan and Katz tend to narrow the distance between the indicative and the subjunctive, thereby legitimating the powers and authorities outside the civil sphere. Mediatized public crises...tend to increase the distance between the indicative and the subjunctive, thereby giving to civil society its greatest power for social change. In these situations, the media create public narratives that emphasize not only the tragic distance between is and ought but the possibility of historically overcoming it. Such narratives prescribe struggles to make "real" institutional relationships more consistent with the normative standards of the utopian civil society discourse. (Alexander & Jacobs, 1998, p. 28)

This approach revises traditional Durkheimian views of ritual as necessarily binding collectivity and also distances itself from neo-Marxist outlooks predisposed to view all public rituals as deterministically and materialistically working in the service of hegemonic interests. Here the analytical focus shifts to the performative, processual and contingent nature of mediatized public crises and how the latent power of civil societies, rooted in deep-seated cultural presuppositions and normative horizons, can become actualized within them (Alexander, 2006a). Once the mediatized wheel of a public crisis begins to turn, public performances, it seems, are obligated, cultural scripts become resurrected, symbols are deployed and "performing the binaries" gives shape, form and cultural meaning to the myths and discourses of civil society. Approached thus, mediatized public crises, from Watergate to the political fallout from September 11, become theorized and conceptually explicated in terms of "cultural pragmatics," "symbolic action" and "ritual" (Alexander, 1988, 2009; Alexander & Smith, 2003; Alexander, Giesen, & Mast, 2006)—ideas that also have traction on potentially disruptive and disintegrative events including disasters. These ideas deserve some unpacking.

For Alexander and his colleagues working within the "strong programme in cultural sociology" (Alexander et al., 2006; Alexander & Smith, 2003), public performances only become effective when they "reinvigorate collective codes" (Alexander, 2006b, p. 80). It is in these "fused" moments that the civil sphere comes alive, publicly instantiated in and through the communication of its utopian discourses (of how society *should* be) and its appeals to solidarity and elaboration of wider "structures of feeling." Cultural pragmatics, theorized as strategic or instrumental action that is necessarily conducted through wider cultural codes, symbolism and meanings, is both conditioned by but also serves to instantiate something of the civil sphere when operating, as it must, "between ritual and strategy" (Alexander, 2006a, 2006b, 2009). Here, then, the civil sphere is not conceived in Habermasian (Habermas, 1989) terms as a rational "public sphere" of deliberation that is allergic to the cultural, symbolic and performative but rather as the widely diffused and deep cultural values and myths that coalesce under the normative horizons of civil society. These are available in all societies and become periodically mobilized in mediatized public crises—whether advanced on the basis of ideas and values of "justice," "fairness," "democracy," the "good society" or, indeed, their binary opposites when ascribed to the "enemies" of civil society.

Mediatized public crises, including disasters that spill over into disagreement and disruption, are not the only ways in which disasters become publicly enacted in the media and are made to mean. Also drawing on the canon of ritual studies and going culturally deeper into how disasters appeal to collective bonds and solidarities, we can see that the original formulation of "media events" (Dayan & Katz, 1992) in fact still has relevance for an understanding of some mediated disasters when constructed and performed in ways that seek to inscribe collectivity and invoke bonds of solidarity. Though the event of a disaster may be unplanned and unscripted (though not as we have heard necessarily unforeseen or unpredictable), successive forms of political response and public engagement can in fact involve high degrees of preplanning (disaster appeals, memorials, remembrance days) and deploy cultural scripts that, by definition, remain available for such occasions. As this brief excursive has sought to highlight, ideas of media performance and media ritualization register in the theorization and conceptualization of disruptive media events and mediatized public crises, including disasters, and can help us better understand how disasters are made to mean and politically matter.

Conclusion

From the foregoing a number of key findings can be highlighted which inform our general approach to mediated disasters in a global age. Though some disasters, emergencies, accidents and crises continue to occur seemingly haphazardly and without warning in particular local-national settings, increasingly many can no longer be assumed to be discrete "disruptive events," sealed behind national borders and emanating from settings localized in time and space. In an interconnected, globalized and mediated world, disasters are often best conceptualized and theorized in relation to endemic and potentially encompassing global crises that are themselves expressive of late modernity and contemporary globalized (dis)order (Cottle, 2011a). Nonetheless, disasters and crises both "old" and "new" have all become increasingly dependent on media and communications in respect of how they become known and responded to. Disasters and crises today are principally defined, dramatized and constituted in and through media and communications.

There is considerable complexity at work, however, in the media's different constructions of disaster and how these register social relations, cultural meanings and political power as well as processes of global interdependency. Mediated disasters can variously be theorized as opportunities for the legitimation of political authority and economic power as well as occasions of critical reflexivity in which political projects, contending discourses and the voices of dissent seek to mobilize and build support for their cause. A new cacophony of voices and views can also now circulate and infuse disaster communications, launched through increasingly available and interlocking communication networks, helping emergency services to focus their efforts and resources and challenging erroneous official claims or ineptitude. Mediated disasters can also be performatively enacted and culturally charged by the news media, inviting and instantiating a moral universe in which boundaries of community, from the local and national and the international to transnational, are variously redrawn and bonds of solidarity correspondingly invoked. But to what extent and in what way exactly disasters and catastrophes may serve as "cosmopolitan moments," based in part on the "globalization of emotions" (Beck, 2006, 2009), cannot simply be assumed.

The scale of death and destruction and the potentially catastrophic results of major threats and disasters, we know, are no guarantee that they will necessarily register prominently in the world's news media. So-called "forgotten disasters," "hidden wars" and "permanent emergencies" still abound in the world today and, because of their media invisibility, often command neither wider recognition nor political response. But we have also heard how the news media are capable on occasion of staging and narrating crises in ways that serve to invest them with emotional and cultural resonance, forcing them into the public eye and appealing to identities of moral community and political ideas of the public good. This chapter, then, has served to point to a more variegated and expressive range of media responses to disaster than is often countenanced by established social science approaches, whether framed through disaster communication specialists preoccupied with processes of disaster cognition and information flows or indeed more critical accounts focused on the political elite-media nexus. This differentiation and added complexity flow, in part, from the news media's occupation of cultural space in the contours and contests of civil society, its professionalized obligations to the institutions and processes of governance, as well as its own logics and autonomous pursuit of corporate and competitive goals in a rapidly changing communication environment. The chapters that follow pursue further these dimensions of difference and dynamics of change while underlining the central and constituting role of media and communications in a global world of proliferating disasters.

Note

1. There are, of course, many different ways of conceiving and theorizing "global crises" today (Cottle, 2011a). Interstate political rivalries and the world's current political "trouble spots" have often been construed through an international relations prism as "global crises," especially when threatening to escalate into regional conflicts and/or embroiling multiple state powers. But global crises today, as discussed, are not delimited to or always best conceptualized in such terms (see, e.g., Abbott et al., 2006; Ahmed, 2010; Amnesty International, 2009; Boyd-Barrett, 2005; Energy Watch Group, 2007; Glenn & Gordon, 2007; Held, 2004; Held, Kaldor, & Quah, 2010; International Institute of Strategic Studies, 2007; Lomborg, 2009; Lull, 2007; Oxfam, 2009a, 2009b; Seitz, 2008; Shaw, 1996; United Nations, 2009; United Nations Environmental Program, 2007; World Health Organization, 2007).

3

The Geopolitics of Disaster Coverage

IN AN INCREASINGLY INTERCONNECTED WORLD, SCHOLARS ARE PAYING ATTENTION to the globalization of mass media and the public sphere and, in particular, the ways in which disasters and conflicts play out on a global stage. As Chapter 2 demonstrated, disasters both crystallize and dramatize conflicts and tensions of political import in our globalized society, and the media play a central role in setting the stage for this drama. Despite these globalizing tendencies, media also frequently frame and anchor our understanding of events within national contexts. They construct narratives that allow citizens to make sense of disasters within the framework of the nation state and its relationship to global power relations, what we here refer to as the *geopolitics of disaster coverage.*

This chapter focuses on the interplay between journalistic practices and the construction of citizenship in disasters. Disaster coverage, like other forms of journalism, constructs citizens at several different levels: First, it makes citizens out of *spectators*, by reminding them of their place in the world and their responsibility to distant others. Second, it makes citizens out of the *victims* of disasters, placing them in the context of their "home nation" and, we argue, occasionally providing them with a political voice. This chapter examines how these forms of citizenship are constructed and how such constructions vary according to a complex confluence of ideology and professional practice.

Journalistic news selection, shaped both by geopolitically inflected ideologies of the nation state and engrained news production processes, is grounded in ideas of proximity and identification which invite audiences/citizens to see themselves and their responsibilities in relation to ongoing global political dramas. We examine how such ideological presuppositions spill over into the language of journalism, through "banal nationalism" and conventionalized constructions of distant victims of disaster. Finally, we consider how journalistic storytelling practices position victims, offering opportunities for political voice which come at a high price. Ultimately, we conclude that the forms of citizenship constructed through disaster reporting tend to reinforce the nation state as the locus of action rather than necessarily promoting cosmopolitan sensibilities.

Journalism, Citizenship and Democracy

The idea that journalists are servants of the public, and that journalism therefore plays a key role in democracy, is central to journalistic self-understandings. Media operate as central institutions of the public sphere, enabling communication between citizens and political leaders. McNair (2007) has summarized the key roles of media in ideal-type liberal democratic societies as follows:

> First, they must *inform* citizens of what is happening around them....Second, they must *educate* as to the meaning and significance of the "facts."...Third, the media must provide a *platform* for public political discourse....The media's fourth function is to give *publicity* to governmental and political institutions—the "watchdog" role of journalism....Finally, the media in democratic societies serve as a channel for the *advocacy* of political viewpoints. (pp. 18–20)

As we illustrate in more detail in later chapters, all of these responsibilities are enacted through media coverage of disasters. And though understandings of these roles have been hotly contested and critiqued, they remain foundational to the self-understanding of professionalized journalists in most advanced societies (see Weaver, 1997; Hanitzsch et al., 2011). However, there is an inherent vagueness to the description of these roles which covers over a series of questions fundamental to journalistic work: How do journalists decide which events merit coverage? How do they determine which facts are meaningful in the first place? How do they judge who should be given a platform for discourse and advocacy and what kind of platform it should be? What are the limits and possibilities of their scrutiny of institutions? What is clear is that the dominant liberal democratic model is firmly planted within the framework of governance structures of the nation state and that the prescribed journalistic roles are articulated in relation to the nation state and its citizens.

In the often-quoted definition by T. H. Marshall, citizenship "is a status bestowed upon those who are full members of a community. All who possess the status are equal with respect to the rights and duties with which the status is endowed" (1950, pp. 28–29). On the basis of a liberal democratic framework, the citizens served by the news media are the subjects of the community of the nation state. Marshall's definition is both pleasingly simple and ridden with conceptual headaches. For instance, do we understand citizenship in the classical liberal democratic sense, as pertaining to the rights and duties of nation state subjects? That is to say, is citizenship primarily enacted through voting in elections and keeping up to date with political information and rewarded by formal civil rights and liberties? (Wahl-Jorgensen, 2007). Or can we talk about it in terms that draw on much broader conceptions of community membership, as when scholars discuss "cultural citizenship" (e.g., van Zoonen, 2005), bringing to bear an increasing awareness of rights and liberties on a wider set of realms and freeing the concept from the tethering of the nation state? Certainly, the nation-centered view of citizenship contrasts with a growing body of scholarship that seeks to understand how globalization affects governance and subjectivity. As we briefly discussed already, scholars have argued for the importance of "cosmopolitan citizenship" which recognizes that "separate states and other actors have an obligation to give institutional expression to the idea of a universal communication community which reflects the heterogeneous character of international society" (Linklater, 1998, p. 23). Beck has been particularly influential in locating the need for a cosmopolitan imagination in the context of a "global risk society" where environmental problems, terrorism, wars and disasters require new visions of politics which go beyond the nation state, even if they are located in decidedly localized events and risks. These are what he referred to as "glocal" questions. As Beck (2000) argued:

> The underlying basis here is an understanding that the central human worries are "world" problems, and not only because in their origins and consequences they have outgrown the national schema of politics. They are also "world" problems in their very concreteness, in their very location here and now in this town, or this political organization. (p. 15)

Beck, however, recognized that the nation state remains the regulatory and institutional locus of political activity while the idea of a cosmopolitan democracy is, for the time being, "no more than a necessary utopia" (Beck, 2000, p. 14). To Beck (2002), realizing a cosmopolitan vision "is about freeing it from national tunnel vision and connecting it with an openness toward world interests (http://logosonline.home.igc.org/beck.htm).

This chapter seeks to uncover some of the complexities of the dialectic of the local, national and global in relationships among citizens, nations and media organizations, viewed through the lens of disaster coverage. It is, of course, impossible to generalize about the priorities, practices and discourses of media organizations: The term covers a variety of different types of media—print, radio, television, online—operating in an increasingly globalized context. Likewise, the audiences and citizens to whom they address themselves vary according to whether they are global, national, regional or local in reach. This means that there are exceptions to the nation-centered approach outlined earlier—major international broadcasters, including CNN, Eurovision, Al-Jazeera and the BBC, address themselves to a global audience and the forms of citizenship that they construct could be argued to revolve in part around ideals of universality and cosmopolitanism. Nevertheless, the majority of media organizations operate in far more localized environments and markets. Through the stories they cover, the angles they take on these stories, and the editorial opinions offered to assess them, they construct our shared meanings: Citizens come to know what counts as important information, and they also acquire an understanding of the available and meaningful responses. Media coverage, as Chouliaraki (2008a) puts it, delineates "conditions of possibility for public action" (p. 832). By contrast, distant occurrences that are *not* covered by the media might as well not exist—absence and silence render them invisible and hence beyond the scope of possible action (see, e.g., Moeller, 1999, Chapter 1). What is clear is that the media are central to the construction of citizens in what is primarily a national, secondarily a global, context.

As we discussed in Chapter 2, disaster coverage pays less attention to the fact that disasters are often predictable and predicated on systemic and political problems that have long been known, even if they have rarely been the subject of media coverage. In this respect, conventional understandings of media responsibility clarify views of exactly what it takes for an event or occurrence to step over the threshold of newsworthiness (cf. Sorenson, 1991). In the context of journalistic production processes around disaster coverage, arguments about what audiences need to know, and how this information should be presented, are shaped by justifications based on the centrality of the nation state.

The relationship between the nation state and mass media has long been a central concern of media scholars. Jürgen Habermas' (1989) description of the public sphere as the space between the state and civil society where individual citizens come together to form a public independent from and often against state authority relied on the nation state as its ontological framework. His historical account traced the emergence of print journalism in England, France and Germany in the 17th and 18th centuries as integral to the birth of the public sphere: It was print publications, such as pamphlets and newsletters, which facilitated the shared discussion among individuals and groups in different locales. To Benedict Anderson (1991), the rise of print journalism made the nation state possible. By reading newspapers, he proposed, individual citizens come to see themselves as members of "imagined communities": individual members will "never know most of their fellow members, meet them, or hear of them, yet in the mind of each lives the image of their communion" (p. 15). On this basis, it is perhaps not surprising that "the nation" does not merely have a physical manifestation in the form of government, its representatives and its laws but also, and perhaps more importantly, an ideological manifestation in discursive formations of nationalism which pervade the everyday lives of citizens. Michael Billig (1995) coined the term "banal nationalism" to describe how "routinely familiar habits of language" act as constant reminders of nationhood (p. 93). By this, he meant that public discourse continuously "points" to nationhood. Rather than being an abstraction distant from the lives of most citizens, the nation "is near the surface" of everyday existence. The "ideological habit" of the continual invocation of the nation enables "the established nations of the West to be reproduced" (p. 6). Through this invocation, the ideology of the nation state becomes naturalized. Among other things, Billig was interested in how the language of journalism subtly but continuously "points" to the nation, whether through weather reports, sports journalism or foreign news reporting. He suggested that any glance at a newspaper serves as a reminder of one's belonging within a national community, and cited selected headlines from one day's newspaper articles, including "Britain basked in 79° temperatures yesterday," "Britain's latest cult heroes" and "Is the British teenager dead or just resting a bit?" (Billig, 1995, p. 112).

Billig's framework shows the inescapable power of banal nationalism, insofar as the ideology of the nation state pervades even the most seemingly innocuous mediated discourses. Billig was critical of scholars of globalization as well as what he described as proponents of "postmodern" approaches who espouse the demise of the nation state. In drawing on American discourses in an international context, he argued that even globalized media narratives proclaim a universalized form of nationalism which advances the superiority of the "home" nation.

Here, we suggest that in the context of disaster coverage in particular, globalization means that although media coverage invariably orients itself toward the nation state through the mechanisms of banal nationalism, it is not simply the case that media coverage is inward-looking or neglects other nations. Instead, globalized power relations shape the view from the nation state. As we discussed previously, because of the global nature of contemporary disasters, an oil spill off the coast of Mexico, a flood in Pakistan or an earthquake in Japan has implications that reverberate around the world. The ways in which these disasters are discursively constructed reveal anxieties about the place of the nation in the world, as these constructions play out dramas ranging from the war on terrorism; the relationships among "major players," such as the United States, China and Europe; and histories of colonialism and conflict. It is inscribed with metanarratives that tell "us" how to relate to everyone else.

Scholars have long noted that whenever terms like "we" and "us" are deployed, such language does, at the same time, assume a "they" and "them," and while such distinctions may be based on any form of difference, media discourses highlight the centrality of national identity and citizenship (e.g., Billig, 1995). As Julia Kristeva (1991, p. 96) wrote, "the foreigner is the one who does not belong to the state in which we are, the one who does not have the same nationality." The structured discourse of news creates a polarization between the ingroup and the outgroup (van Dijk, 2009, p. 199), representing "them" as different in order to secure and bolster "our" own identities and, in doing so, reveal much about how we see ourselves. However, this is not to suggest that anyone outside the "ingroup" (Tajfel, 1981) is merely viewed as an undifferentiated "foreigner" or "Other." Rather, as we argue elsewhere in the book, through discursive representations in disaster stories—of who the victims are, where they live, what they say and leave unsaid, what they wear, what they are doing, what others say about them and on behalf of them, among many other things—journalism creates a differentiated picture of the world which signals to audiences how these others should be viewed and what "our" responsibilities are to "them."

Here, we seek to politicize constructions of disasters by paying attention to what we call the "geopolitics of disaster coverage." By this term we mean to emphasize that under conditions of globalization, where the media have become an arena for the ongoing reinforcement and contestation of global power relations, disaster coverage and the forms of citizenship it constructs are heavily influenced by geopolitics at the levels of ideology and journalistic practice. As such, this term enables us to understand the subtle ways in which disaster coverage is inherently political as it always points to the nation state and its role within dynamic global power relations. Broadly speaking, geopolitics, as a field of scholarship, attempts to understand and problematize relations between and among regions, nations, groups and cultures, shaped as they are by complex histories of collaboration, conflict and colonialism. A geopolitical approach takes an interest in ideological constructions of global power through discursive formations, engaging "the geographical representations and practices that produce the spaces of world politics" (Ó Tuathail & Dalby, 1998, p. 2):

> Critical geopolitics is particularly interested in analyzing the interdigitation of all these practices, in examining how certain conceptual spatializations of identity, nationhood and danger manifest themselves across the landscapes of states and how certain political, social and physical geographies in turn enframe and incite certain conceptual, moral and/or aesthetic understandings of self and other, security and danger, proximity and distance, indifference and responsibility. (Ó Tuathail & Dalby, 1998, p. 4)

We borrow from critical geopolitics the idea that hegemonic discourses and representations reflect complex and dynamic global power relations, rather than simply orienting themselves to the nation state. This, in turn, signals the need for attention to the ways in which disaster coverage constructs relations between and among countries and regions and their citizens. As Chouliaraki (2006) argued, the relationship between spectators and the distant victims of disasters is inherently geopolitical, reflecting thus the global distribution of power: "...it is always 'audiences in North America or Western Europe [that] react to knowledge of atrocities in East Timor, Uganda, or Guatemala,' rather than the other way round" (p. 22). Geopolitics plays a role in shaping journalistic practice because it informs the allocation of resources and the prioritization of particular stories over others. When disasters make the news, it is often the result not merely of any objective measure of the "seriousness" of the disaster, as indicated by the loss of life, but of *where* the disaster happened and *who* it affects (see also Mody, 2010). This, in turn, has profound consequences for how audiences understand the world.

News Values and Geopolitics

Some scholars have suggested that what Galtung and Ruge (1965) first described as the "threshold" of an event plays a role in determining which disasters make the news: The more casualties there are, the greater the likelihood that the event will be covered (Gaddy & Tanjong, 1986; Simon, 1997). Certainly, the event orientation of disaster news focuses our minds on spectacular events leading to extensive loss of life (such as an earthquake), to the detriment of examining ongoing, "everyday disasters" (such as environmental pollution and car crashes; cf. Moeller, 1999).

However, matters of geopolitics appear to be equally crucial in shaping the priorities of disaster journalism. A 1995 large-scale content analysis by the Pew Research Center for the People and the Press concluded that "the way the media covers international news may be doing little to change the American public's indifference to concerns about world events and foreign policy" (p. 1). The study highlighted features of international coverage which have been consistently documented by scholars: News organizations tend to prioritize domestic news or news that has a clear domestic "angle"; they cover different regions according to their perceived importance and focus on dramatic, violent and conflictual stories. They therefore fundamentally construct "The Foreign" as a place of violence, conflict, danger and spectacle, and disaster coverage is no exception. Although the report identified what its authors saw as distinctive features of American international coverage, such as an emphasis on unbiased and objective reporting, the nation-centered orientation of international news that it identified reflects a broader set of long-standing journalistic practices which cut across geographical boundaries and arise on the basis of a complex combination of ideological causes and professional routines that structure understandings of what counts as "newsworthy." Galtung and Ruge (1965) took up the apparently simple question of how events become news, outlining a set of "news values" informing journalists' selection of stories and angles. Their concept of news values has perhaps rightly been criticized for its "somewhat mythical" character (Richardson, 2005, p. 173), in part because journalists would not choose to articulate their news selection in such terms but would rather describe it by invoking an innate "news sense" or "nose for news" (Molotch & Lester, 1974). But certainly, news values form part of a tacit knowledge which is "passed down to new generations of journalists through a process of training and socialization" (Harrison, 2006, p. 153; Harcup & O'Neill, 2001), an unspoken set of rules which is learned on the job and becomes central to journalists' ways of life.

Galtung and Ruge (1965), in identifying 12 factors that they saw as central to the selection of which events to cover, suggested that stories which take place in proximity to the home nation, which include reference to elite nations or elite people, and which are culturally familiar, are much more likely to make it into the news. Writing at the height of the Cold War and toward the end of the colonial era, they were conscious of the emerging realities of global interconnectedness and critical of how globalization reinforces existing power relations. They took note of the fact that countries of a "lower international rank" (p. 83) are covered in ways reflecting audiences' presumed "mental pre-image" of that nation—while "topdog" countries have no difficulties in attracting nuanced and positive media coverage, "underdog" countries are primarily represented through narratives that focus on negativity and conflict and draw on well-established stereotypical shorthand:

> The consequence of all this is an image of the world that gives little autonomy to the periphery but sees it as mainly existing for the sake of the center—for good or for bad—as a real periphery to the center of the world. This may also tend to amplify more than at times might seem justified the image of the world's relatedness. Everything's relevance for everything else, particularly for us, is overplayed. Its relevance to itself disappears. (Galtung & Ruge, 1965, p. 84).

In this lament, Galtung and Ruge (1965) pinpointed the ideological consequences of journalism's geopolitics: On the basis of mediated constructions of the world, citizens of Western "topdog" countries come to see world events through the spectrum of their nation state and its interests, with severe consequences for the "underdog countries." Other scholars have backed up these central themes, highlighting the role of geopolitics in foreign news coverage (e.g., Harcup & O'Neill, 2001; O'Neill & Harcup, 2009; Mody, 2010), and demonstrated that despite variations over time, the fundamental priorities reflected in the original study remain relatively stable. Elite nations and individuals tend to top the news agenda, and at least in the dominant news cultures of the West, the farther events unfold from the "centre," the less likely they are to be reported.

Peter Golding and Philip Elliott (1999), in their groundbreaking study, *Making the News,* originally published in 1979, took a critical view of the news values literature. To them, the approach is guilty of embracing the mythology of journalism, celebrating the novelty, excitement and creativity of covering unknown and breaking news stories:

> [N]ews production is rarely the active application of decisions of rejection or promotion to highly varied and extensive material. On the contrary, it is for the most part the passive exercise of routine and highly regulated procedures in the task of selecting from already limited supplies of information. (Golding & Elliott, 1999, p. 118)

They argued that decisions about news selection derive from assumptions on the basis of three principal concerns. While the first of these relates to the importance of the story to the *audience,* the second and third factors purely pertain to the ways in which the story fits the practical concerns of the news organization: *Accessibility* is understood in terms of two factors, prominence and ease of capture. Prominence refers to the extent to which the event is "known to the organization," while ease of capture reflects "how available to journalists is the event, is it physically accessible, manageable technically, in a form amenable to journalism, is it ready-prepared for easy coverage, will it require great resources to obtain" (Golding & Elliott, 1999, p. 119). Fit reflects whether the item is "consonant with the pragmatics of production routines, is it commensurate with technical and organizational possibilities, is it homologous with the exigencies and constraints in programme making and the limitations of the medium? Does it make sense in terms of what is already known about the subject?"(Golding & Elliott, 1999, p. 119).

Given that disaster coverage, perhaps more so than most other forms of journalism, is based on unexpected events that frequently take place in what are, for Western media organizations, distant and difficult-to-reach locations, it is strongly shaped by pragmatic decisions. Such editorial decisions do not merely reflect the significance of the event but are also based on a consideration of limited resources, in the context of a highly routinized news production process. This has led some scholars to suggest that news production processes closely resemble factory assembly lines, to the point that we can examine the workings of the "news factory" (e.g., Bantz, McCorkle, & Baade, 1980). Certainly, much journalistic work is based on careful temporal planning designed to limit the uncertainty of producing news. It could be argued that journalism is often marked by a "stop-watch culture" where what is valued is not so much creativity but efficiency and the ability to produce news on a tight schedule (Schlesinger, 1987). Ekström (2002) commented that journalistic "knowledge production is steered by the demands of predictability and control over the ingredients of a programme that has a predetermined format, that will be ready at a given point in time and that recurs with a certain regularity (often daily)" (p. 269).

If temporal aspects of news production dictate many journalistic practices, this is also true of disaster coverage. The pressures of deadlines—only strengthened in recent years by the emergence of constantly updated online news and the rise of 24-hour news coverage—enforce the immediacy-orientation of journalism and militate against contextual stories. Deadline pressures mean that journalists rely heavily on a limited number of easy-to-reach authoritative sources. Tuchman (1978, pp. 21–25) developed the metaphor of the "news net" to describe the system of beat reporting which catches the "big fish"—or the sources who have privileged access due to location and status—but lets the tales of less privileged people and groups slip through the holes. She suggested that the finer the mesh of the web, the more fish that are caught. In the context of disaster coverage, the workings of the news net further reinforce existing power relations. With a declining budget for foreign correspondents as a result of belt-tightening at news organizations around the world, the "news net" is increasingly focused on a few "centers" and increasingly unable to offer reports on places at the margins. The availability of technologies and human resources in particular areas therefore significantly influences whether and how an event will be covered—as Moeller (1999) demonstrated, a long-standing famine in Somalia suddenly came onto the news agenda in the United States after a high-profile military intervention necessitated the deployment of news organization resources—as one foreign editor commented, it was suddenly "on the radar screen in Washington" (p. 126).

By relying on official and powerful sources, and information from easily accessible locations, journalists are not always able to access "facts on the ground" and may perpetuate dominant Western narratives about the causes of the disaster (Lappè, Collins, & Rosset, 1988). Bennett and Daniel (2002) interviewed British journalists covering disasters and found that they relied primarily on UN agencies, disaster relief charities and UK government agencies, as well as reports from other media organizations. In fact, according to Durham (2008, p. 95), the unusually politicized and critical coverage of Hurricane Katrina was a result of the absence of such official sources because of the "federal government's de facto absence" for the first week following the disaster.

The use of authoritative sources is a particular concern in disaster stories, where they represent the "home nation" of the news organization, rather than the disaster-struck region. Such sources may be easy to reach for comment, but they may not necessarily have sufficient knowledge of conditions on the ground. This is all the more problematic because of the increasing use of generalist reporters who are often "parachuted" in to report on events which they may not have the expertise to adequately interpret or contextualize and therefore provide audiences "with a more confusing, disjointed and violent image of the world, rather than an informed and balanced understanding of international events" (Norris, 1995, p. 367). Pedelty (1995) concluded, on the basis of an ethnographic study of foreign correspondents in El Salvador, that parachute journalism is both derivative and ethnocentric because its practitioners "have a greater tendency to return to their own cultural values and social conditions when interpreting the world" (p. 110).

The ways in which the ideological framework of the nation state shapes journalism practice has profound consequences for news coverage. Not only are marginal countries less covered in the West, but their news stories are also less likely to circulate globally. For example, Jaap van Ginneken (1998) found that countries with 10% of the world's population—primarily in Europe and the United States—are the base for 90% of transcontinental flows of media material. Over the period of Akiba Cohen and his colleagues' (Cohen, Levy, Roeh, & Gurevitch, 1996) study of the European news exchange, the service provided almost 20,000 news items to news services in Asia, Africa and the Caribbean, but it took only 555 stories from them in exchange. In other words, the movement of news and information reflects global power relations underpinning news selection.

These power relations also pervade the reporting of disaster, demonstrating "glaring differences that reflect global hierarchies of place and human life" (Joye, 2009, p. 45). The global hierarchy of suffering invests more significance in the disasters and traumatic events that claim Euro-American victims, while suffering outside the West is routinely minimized or ignored. This practice, in turn, normalizes and perpetuates global inequalities. For example, Joye (2010) found that Flemish television news devoted about 50% of its coverage to European disasters, even though they only accounted for a total of 11.4% of all disasters that occurred over the period studied. On the basis of this geopolitical hierarchization, he argued that proximity, understood as a metaconcept covering "cultural affinity, historical links, geographical distance, trade or economic relations as well as psychological or emotional distance" can be seen as a powerful predictor of disaster coverage.

Indeed, measures of proximity in this broader and more politicized sense matter because journalists, in selecting what stories to cover, make assessments about the affective response on the part of the audience. In Winfried Schulz' (1982) work on news selection, one of six main dimensions he identified referred to the audience's ability to relate to the content, based on proximity, ethnocentrism, personalization and emotions. Geopolitics is thus intertwined with journalistic expectations of audiences' affective response. Later chapters go into more detail regarding the relationship between emotion and disaster coverage, but here we should note that scholars are paying increasing attention to the affective elements of political engagement and citizenship. And while liberal democratic theory holds that citizens should be provided with information about ongoing events, epistemologically and normatively underpinned by the ideal of objectivity (cf. McNair, 2007), it is precisely the affective, impassioned responses that bring readers, viewers and listeners into being as citizens who care about distant disasters. As Marcus, Neuman, and MacKuen (2000, p. 1) suggested, "emotion and reason interact to produce a thoughtful and attentive citizenry." Indeed, scholars who analyze disaster coverage through the specters of the "spectatorship of suffering" (Chouliaraki, 2006) and "distant suffering" (Boltanski, 1999) are interested in how the elicitation of emotional response relates to opportunities for cosmopolitan citizenship. As Chouliaraki (2006) has convincingly argued, however, the cosmopolitan imagination is restricted by the nationally and geopolitically inflected narratives of disasters.

As such, media discourses first of all construct their own audiences as citizens of the nation state and the world. Chouliaraki (2006, p. 200) argued that emergency "news is world-making…in so far as its representations of suffering push the spectators' sense of belonging beyond its existing boundaries and seek to constitute their relationship to distant suffering in terms of a demand for public action." At the same time, these representations provide interpretive frameworks for categorizing the citizens of other countries, especially those who are the victims of disaster. Disaster coverage, in turn, offers carefully structured opportunities for the victims to speak as citizens, thus mediatizing an ideologically overdetermined dialogue between those citizens who enter the public eye as a result of their involvement in a major news event and those citizens who are the spectators of the event. Like any relationship forged out of the mediated gaze, it is an unequal one, and the power dynamics of the disaster gaze are informed by geopolitics.

In other words, if there is no such thing as "news from nowhere" (Epstein, 1973), discourses ostensibly generating cosmopolitan citizenship are rarely politically innocent. If respect for the "rights of others" (Benhabib, 2004) is central to cosmopolitan citizenship, the geopoliticized nature of discourses on disaster means that some victims have more rights than others, because their representation is more likely to give rise to compassion, action and intervention. For one, proximity is central to the creation of compassion and solidarity (Boltanski, 1999). Luc Boltanski (1999) and Chouliaraki (2006) have demonstrated how media coverage does not simply represent events but also demands a response in the spectator. As Boltanski (1999) wrote, in "relation to the media, the spectator occupies the position of someone to whom a proposal of commitment is made" (p. 149). The journalist "conveys statements and images to a spectator who may take them up and, through his words, pass on in turn what he has taken from these statements and images and the emotions raised in him" (Boltanski, 1999, p. 149). Boltanski (1999) thus placed the spectator in an active position as someone who can either accept or reject the proposal.

Some have argued that this implicit proposal for action is increasingly rejected. Susan Moeller (1999) has written compellingly about the phenomenon of "compassion fatigue" or the public's limited attention to major global crises, including disease, famine, war and death, caused by what she sees as irresponsibly event-oriented, sensationalist and short-term media coverage which exhausts audiences in its relentless and unending parading of death and horror. Moeller's attack on media coverage has found resonance among scholars, journalists and aid workers. For example, according to one study, charity fundraising managers felt that the media "routinely sensationalized foreign catastrophes and projected highly demeaning images of victims" (Bennett & Daniel, 2002, p. 40). One respondent complained that every "disaster brings the same request for a 'granny up a tree' shot—recreated if need be" (Bennett & Daniel, 2002, p. 40). The thesis of compassion fatigue has nevertheless drawn criticism from various camps (e.g., Chouliaraki, 2006, pp. 112–114). Among other things, critics, while not necessarily disagreeing with Moeller's broad characterization of journalistic practice, have suggested that it fails to acknowledge the myriad representational forms of journalism and the fact that not all disaster coverage positions spectators uniformly. Rather, as we explore later in more detail, there are complex patterns of expression entitlement and forms of actions, determining the extent to which particular representations solicit compassion, identification or "suffering without pity" (Chouliaraki, 2006).

In the introduction, we mentioned the "calculus of death" as a geopoliticized measure of newsworthiness, so embedded into the fabric of practice that journalists knowingly joke about it and have developed insider jargon to describe it. Schlesinger (1987) discussed "McLurg's Law," which establishes a ratio between the size of an event and its distance from (or relevance to) the news audience: "One European is worth 28 Chinese, or perhaps 2 Welsh miners worth one thousand Pakistanis" (p. 118). Moeller (1999) similarly suggested that "there are several tongue-in-cheek equations floating around that purport to formalize the business of deciding what crisis to cover" (p. 21). She cited a journalist enumerating the following considerations:

> Is it a place Americans know about? Travel to? Have relatives in? Have business in? Is the military going there? You're not going to get on page one with something about Bangladesh nearly as much as you do with something about some country where your readers have some kind of connection. (Moeller, 1999, p. 21)

Journalists' common sense practices, based on audience considerations informed by geopolitics, also spill over into journalistic storytelling. What Clausen (2004) wrote about the practices of Japanese and Danish foreign news journalists is an insightful description of the paradox that faces all news workers involved in the production of news about disaster. She observed that "the Janus-faced ability of both knowing international affairs and knowing the receiving audience was found to be essential in the framing of international news information and an important element in the process of presenting events to a national audience" (p. 25). In a globalized media environment, journalists need to possess a complex set of skills: Not only do they have to mediate between the citizens and politicians within the nation state, but they also have to know enough about the needs and knowledge of their audiences to link up citizens of a nation to complex international and global events. Journalists are domesticating global events in rendering them meaningful to national audiences:

> Media maintain both global and culturally specific orientations—such as by casting faraway events in frameworks that render these events comprehensible, appealing and rele-

> vant to domestic audiences; and second, by constructing the meanings of these events in ways that are compatible with the culture and the dominant ideology of societies they serve. (Gurevitch, Levy, & Roeh, 1991, p. 206)

In other words, journalists do their work informed by a complex set of negotiations to ensure the relevance of their news stories to audiences. Domestication is a key journalistic practice across all forms of international news, and disaster reporting is no exception. Stories can be made relevant to national interests through such practices as interviewing each country's own politicians and giving more extensive coverage to "proximate" individuals and stories or through a focus on compatriots in the disaster areas and the national relief aid effort in the affected region (Clausen, 2004).

This does not mean that journalists engage in the deliberate advancement of national and nationalist interests or that domestication operates as a countermeasure to globalization or a form of protectionism (Clausen, 2004). Instead, media organizations can succeed only by means of producing culturally compelling and comprehensible material. By engaging in domestication, they provide information that doesn't merely allow citizens to make sense of the world around them but also sets the boundaries of the universe of issues that citizens are supposed to engage with and act upon. The idea of domestication thus highlights the fact that news production does not take place in an ideological vacuum but instead is strongly shaped by the geopoliticized ideology of the nation state, combining with the constraints of news production to limit the universe of available news stories.

Practices of domestication can tell us much about the relationship between the "home nation" and the country affected by the disaster. For example, on August 11, 2010, the main story covering the then-ongoing floods in Pakistan in the British quality newspaper, the *Guardian,* was headlined "Pakistan president visits flooded regions as official response criticized." It thus instantly framed the story in politicized terms, pointing to the (inadequate) actions of the Pakistani nation state. After several paragraphs focusing on the damage caused by the flooding so far, and the relief effort that sought to address it, the article turned to the British response: "British donors have so far given £10.5m to help flood victims, according to the Disasters Emergency Committee, which said the money had helped provide more than 500,000 survivors with emergency care, clean water, food or shelter." This story was typical in sharing information about the British response to highlight the role of national citizens—as contributors and benefactors. Through this form of domestication in disaster stories, citizens of the home nation appear to stand together, united as "British donors," and as nationally constituted actors their donations have concrete consequences in securing the basic needs of beleaguered fellow citizens of the world. Thus, the article draws attention to acts of global citizenship—at least in the vaguest sense of caring for distant victimized others—but in the context of belonging to the nation state.

A *New York Times* article published at the same time followed a similar pattern in first outlining the latest news about the disaster: "Pakistan issued new flood warnings on Thursday that could last into the weekend as government and relief agencies strained to confront the toll from a growing humanitarian disaster" ("New flood warnings raise fears in Pakistan," August 12, 2010).

After discussing President Zardari's visit to affected regions, the article went on to provide more detail about the U.S. response:

> The United States Embassy in Islamabad announced that two Marine helicopters had arrived in the country, the first of a contingent of 19 American military choppers that has been ordered to assist the Pakistani government in relief efforts. The United States has pledged $71 million for flood relief, and American officials have called for more.
>
> "Americans have been very focused on other, equally heart-wrenching, issues, like Haiti," Richard C. Holbrooke, the special representative for Afghanistan and Pakistan, said Wednesday at the Council of Foreign Relations. "I hope they will turn their attention as well to this extraordinary crisis that Pakistan is facing.
>
> The aid deliveries could help the United States improve its image here and blunt a growing anti-American sentiment. The Taliban have already urged the Pakistanis to shun American aid and have used the current crisis to expand their influence and outreach in the flood-affected areas of the northwest. ("New flood warnings raise fears in Pakistan, "August 12, 2010).

This *New York Times* story, in following the template of domestication, also provided details of the distinctive contributions of the "home country" and included an appeal from Richard Holbrooke for further contributions. Holbrooke's statement clearly hailed media audiences as American citizens, reminding them at the same time of their global responsibility. This global responsibility, in turn, was placed in the context of the need to "blunt a growing anti-American sentiment" in the region as a result of the country's engagement in conflicts in Iraq and Afghanistan. As such, the *Times* story about the Pakistan flood was as much about America' s place in the world and the responsibilities of its citizens, both to the suffering masses of Pakistan and to the image of their country abroad.

By contrast, an article published by the Associated Press of Pakistan on August 15 focused on the floods as a historic opportunity to demonstrate the strengths of the nation:

> Prime Minister Syed Yusuf Raza Gilani on Saturday expressed the resolve that the resilient Pakistani nation was strong enough to cope with any challenge and would rehabilitate the millions uprooted by the devastating floods. Addressing the nation on the Independence Day, the prime minister said, "the nation will confront the challenges as it did at time of partition, with the blessings of Almighty Allah. This time too we shall succeed." ("Nation is strong to cope with challenges: PM," August 15, 2010)

This story, drawing on a series of markers of banal nationalism, in the context of the country's independence day, also situated Pakistan within the landscape of global power relations, insofar as it included appeals for the world community to provide further assistance and went into detail about the prime minister's statements on international relations, reporting that he:

> [S]poke on the main contours of [the] country's foreign policy and said it was reflective of national aspirations for peace and stability within Pakistan, in [the] neighbourhood, the region and the world. He said Pakistan desires to develop partnerships of mutual benefit with the entire international community. (Associated Press of Pakistan, "Nation is strong to cope with challenges: PM," August 15, 2010)

Similarly, when Japan was struck by a magnitude 9.0 earthquake that hit the north of the country on March 11, 2011, and a subsequent tsunami that wiped out coastal villages and towns, the disaster coverage, as it unfolded, highlighted the central geopolitical role of Japan. Initial stories followed the pattern of other disaster reporting, focusing on the horror and grief of the events and communicating basic information about the scale of the destruction wreaked by the earthquake and the tsunami. One of the first print stories about the disaster, published in the *Belfast Telegraph,* told of how "TV footage showed waves of muddy waters sweeping over farmland near the city of Sendai, carrying buildings, some on fire, inland as cars attempted to drive away" ("Massive Japan earthquake triggers tsunami; Buildings swept away by 30-foot waves as coastal towns severely damaged," March 11, 2011). Even more graphically, the Australian *Advertiser* reported on the immediate aftermath of the tsunami:

> It was an unstoppable nightmare—a wall of water, mud and man-made missiles that spared nothing. It carried ships and hundreds of houses for kilometres as it raced across cities, towns and farmland and chased down people hopelessly trying to flee by foot, bicycle and car.
>
> Where buildings and bridges didn't give under the weight of the killer wave, debris-laden water was shot high into the air and rained down on buildings.

> Freeways and roads were peeled away by the raging waters, the vehicles caught on them turned into torpedoes as they were smashed into other structures and were tossed around as if in a washing machine. ("Horror strikes Japan," March 12, 2011)

These initial stories were stylistically and substantively generic, insofar as they deployed common tropes shared across all disaster reporting (Pantti & Wahl-Jorgensen, 2007), graphically illustrating the consequences through charged language, including terms like "unstoppable nightmare," "killer wave" and "raging waters"—ones which vividly evoked the undiscriminating force of nature and human powerlessness in the face of it. At this early stage of disaster coverage, geopolitics appeared to have less of a bearing on the narrative than the need to tell the tale of what happened in an evocative manner.

Nonetheless, questions of geopolitics quickly came to the forefront. First of all, much reporting centered on understanding the distinctive nature of Japan and the country's apparently significant resources and capacities for handling the disaster. The coverage emphasized the high level of earthquake preparedness, quality of construction and efficient and stoic response to the disaster. In the two weeks after March 11, 208 stories in major world newspapers used the word "stoic" to describe the Japanese people's reactions. One article, titled "No Donor Rush to Aid Japan," published in *USA Today,* even explained the limited generosity of American citizens with references to Japanese stoicism, which failed to generate empathy:

> Another element is the stoic nature of the Japanese people, who may not evoke as much empathy in the American media audience. A lot of Japanese people don't want to give media interviews out of sense of humility, Rooney said.
>
> "They have this 'could have been worse' attitude, which may be a great mentality and attitude for survival, but it's not great at generating a great deal of interest and philanthropic support," he said. [...]
>
> Finally, it may be the Japanese culture of deep communal loyalty, of fortitude, that also dampens on American giving.
>
> Far from appearing helpless, Japanese stoicism and civility has been remarkable to observers across the world since the earthquake.
>
> It emanates from a culture forged on an earthquake-prone rock in the middle of a stormy ocean.
>
> "Suffering and persevering is a type of virtue in Japan," Stewart says. "They identify with it and are proud of it, the ability to persevere and remain calm under difficult situations." (March 16, 2011)

Another article, published in the *Weekend Australian* and headlined "Apocalypse Shows Truth of the Japanese Soul," backed up the narrative of Western impatience with the unemotional response of the disaster victims, suggesting that "these are a stoic and disciplined people, conditioned from birth to respect authority" but that "an outsider can't help wondering if in the face of this disaster, the people need to show the normal human reactions of grief and anger" (March 19, 2011). The discourse of Japanese stoicism in the face of disaster was repeated so often that it became, in the eyes of some observers, a convenient shorthand covering over a deeply embedded national stereotype. As Tom Huang (2011) wrote in a commentary for the Poynter Institute, in a careful critique of the limitations of this approach, the narrative led 'us to believe that there are cultural roots in the Japanese citizens' stoic response to all the horrors of the past few weeks' (Huang, 2011), whereas in fact, there is evidence to suggest that people around the world will tend to remain calm in the face of trauma and disaster.

As such, the entrenched discourse of the stoic Japanese character became a reflexive way of "Othering" the victims of the devastating disaster, often resorting to essentializing ideas around the self-sufficient island state. This narrative, in turn, drew attention away from the human suffering associated with the earthquake and tsunami and went hand in hand with what became the dominant way of constructing the disaster: Very quickly, much of the media's attention turned to the disaster's consequences for the global economy, shaken by the long-standing financial crisis as well as by upheaval in the Middle East. This coverage highlighted the geopolitical significance of Japan. Amid stories displaying discourses of horror and grief, with tales of dazed survivors searching ruined cities for missing family members, of wrecked ships washed hundreds of yards ashore and vanished cities, many of the stories were of financial woe.

Media reports highlighted the interconnectedness of the global economy, as everything from the sale of smartphones to the production of cars was affected by the earthquake. The shocks of the earthquake, in other words, were felt as shocks to the global economy, as automotive factories around the world had to shut down and the stock markets reeled. For example, a commentary in the British quality newspaper, the *Daily Telegraph,* published on March 14, 2011, three days after the earthquake and tsunami struck, opened by suggesting that "amid the scale of the human catastrophe, any consideration of the economic consequences of the Japanese earthquake might seem faintly distasteful if not even perverse. Yet any disaster on this scale also has the capacity for extreme long-term economic damage and Japan is going to require careful management over the months ahead" ("Tragedy might be final straw for fragile confidence," March 14, 2011). It then considered in detail the financial consequences of the disaster in a global context:

> The immediate impact will almost certainly be to push Japan back into technical recession.
>
> Fortunately, the area most directly affected is quite sparsely populated by Japanese standards and in economic terms not particularly important. The region hit by the Kobe earthquake in 1995 was in this regard of considerably more consequence.
>
> Nonetheless, the effect of the quake has been to bring economic activity to a grinding halt, rather in the way September 11, a comparatively localised event, managed to create a temporary hiatus across the entire US economy. [...]
>
> All the same, the history of such disasters is that after a brief lull, economic activity tends to bounce back, boosted by heavy spending on renewal. (*Daily Telegraph*, "Tragedy might be final straw for fragile confidence," March 14, 2011)

After this assessment, which, among other things, reassuringly stated the relative insignificance of the affected region and could be viewed, by the admission of its author, as both distasteful and perverse, the article concluded as follows: "What can be said with certainty is that the world economy needed another disaster like a hole in the head. The best that can be said for this one is that unlike most economic disasters, at least it wasn't man-made" (*Daily Telegraph*, "Tragedy might be final straw for fragile confidence," March 14, 2011). This piece was not unique in its callousness but instead was symptomatic of a larger trend of focusing attention away from the scale of human suffering and toward questions of the impact of the disaster on global and national economies. The scale of this coverage was unprecedented: A Nexis UK search showed that in the two weeks following the earthquake and tsunami, there were a total of 1,454 articles on the global economic consequences of the disaster in major world newspapers.

When it became clear that the Fukushima nuclear plant, caught up in the tsunami floodwaters, was threatened by a meltdown, the global interconnectedness brought to the forefront by the disaster became an even more urgent topic of media coverage, as countries around the world reconsidered their nuclear energy generation plans, and advocates and opponents of nuclear power engaged in debates over safety with renewed impetus. Most importantly, even if much coverage considered the environmental risks of the nuclear disaster in a global context, the impact on the economy continued to dominate the discussion. Here, an opinion piece in the Singapore *Business Times* considered the disaster through the lens of the close ties between the Asian economies:

> Japan may have fallen behind China (as of last year) as the world's second-largest economy behind the United States, but its continuing importance to the rest of Asia and to the global economy has been dramatically underlined by the impact of the twin disasters that struck the country this month.

> Japan is at the centre of a regional production network and supply chain which in turn keeps much of the rest of the world supplied with finished goods, as (ADB) president Haruhiko Kuroda emphasised in a special interview with *The Business Times* in Tokyo last Friday.
>
> As such, Japan's economic importance continues to be greater than that suggested simply by national output or gross domestic product, he noted.
>
> There are many lessons to be learnt from the disasters that Japan has endured after being hit by a magnitude-9 earthquake and tsunami and then a nuclear power disaster which has surpassed that of the Three Mile Island in the United States, Mr. Kuroda told *BT*.
>
> Among these are the need for Asia as a whole to strengthen its disaster-prevention and response systems and for the region to find ways of reinforcing the production networks and supply chains that are the lifeblood of its economy. (*Singapore Business Times*, "Lessons for Asia from Japan disasters," March 25, 2011)

Clearly, an emphasis on the economic consequences offered yet another means of news domestication. At the same time, it demonstrated the complex workings of the geopolitics of disaster reporting, and specifically, the ways in which concerns for "our" economy—global and national—can trump concern for "their" victims.

It seems that the geopolitics of disaster coverage and the nation-centric nature of journalism affect story selection and the ways in which stories are domesticated (Clausen, 2004) or framed (e.g., Entman, 2004), and shape the language of news in both explicit and subtle ways. The language of disaster coverage, and the discourses on the "Other" that it produces, have long been studied by scholars who have raised larger questions about the citizens produced as spectators and subjects of the texts. As Sorenson (1991) pointed out in his discourse analysis of coverage of the Ethiopian famine, Western media often draw on a stock of essentializing racist discourses in covering disasters and famines in Africa—ones in which the victims' misfortune "is attributed to a type of African essence rather than to structural conditions inherited by the post-colonial states and the specific historical circumstances of Africa's integration into a world-market system" (p. 235). These discourses echo the justifications for colonialism put forth by colonial states and remain in evidence (if in a more marginal and marginalized fashion) in the disaster coverage we examine in this book.

The tropes that are commonly employed in disaster coverage thus both distance and "other" the victims. Along those lines, Gregory Bankoff (2001) described how, in the production of disasters, these tropes are part of the following:

> [T]he same essentialising and generalising cultural discourse: one that denigrates large regions of the world as dangerous—disease-ridden, poverty-stricken and disaster-prone; one that depicts the inhabitants of these regions as inferior—untutored, incapable, victims; and

> that reposes in Western medicine, investment and preventive systems the expertise required to remedy these ills. (p. 29)

Thus, the dominant discourse of disaster positions the dominant Western "donor" countries as superior and the "receiver" Third World countries as backward victims. Such readings are reinforced by coverage of Western relief efforts. One UK charity representative appeared on television to complain about coverage of a flood in Mozambique, accusing the media of showing only "blond haired Aryans swooping from the skies in helicopters to pluck poor little African babies from the treetops" (Bennett & Daniel, 2002, p. 35). Bennett and Daniel (2002) commented as follows:

> Reporting of this nature arguably patronises and belittles victims and helps create negative stereotypes of Third World communities as being confused, bewildered, incompetent, helpless, lacking self-reliance and living in countries with governments that are unable either to mount relief efforts or to plan ahead in order to avoid potential catastrophes. (p. 35)

As such, these discourses do not so much elicit compassion as reproduce existing geopolitical power relations (cf. Benthall, 1993, pp. 186–188). Bankoff (2001) suggested that countries seen as disaster prone are geographically "othered" through a media discourse which both naturalizes and sensationalizes their regions:

> [T[he disproportionate incidence of disasters in the non-Western world is not simply a question of geography. It is also a matter of demographic difference, exacerbated in more recent centuries by the unequal terms of international trade, that renders the inhabitants of less developed countries more likely to die from hazard....No single term has yet emerged that defines the areas where disasters are more commonplace: the media often sensationalises a certain region as a "belt of pain" or a "rim of fire" or a "typhoon alley," while scientific literature makes reference to zones of "seismic or volcanic activity" or to meteorological conditions such as the El Niño....Whatever the term, however, there is an implicit understanding that the place in question is somewhere else and denotes a land and climate that have been endowed with dangerous and life-threatening qualities. (p. 24)

Of course, it is not merely particular disaster-struck regions that are subject to such distancing representations but also the victims (see Silk, 2000). Nevertheless, it is a strong and enduring journalistic convention to provide direct evidence of disaster victims and their suffering, and to do so, reports frequently draw on eyewitnesses and victims and their families as sources. The limited presence of "ordinary citizens" in routine coverage has long been noted by scholars. Galtung and Ruge (1965), in their study of foreign news coverage discussed earlier, documented that "ordinary people" had a relatively low prominence and presence in news coverage.

This finding is not only restricted to international news but also reflected in media coverage more generally. Herbert Gans, in his classic (1980) ethnographic study of news production, found that "knowns" are four times as likely as "unknowns" to appear in the news. Ordinary people would only make an appearance if they were caught up in major news events as victims or eyewitnesses. However, research on how ordinary citizens are represented in the news media, drawing on a content analysis of television and newspaper coverage in the UK and the United States in 2001 and 2001, found that ordinary people and their opinions appeared relatively frequently in the news (Lewis, Inthorn, & Wahl-Jorgensen, 2005). The apparently increasing presence of ordinary people in the news may be a sign of shifting priorities in news organizations in response to audience demands.

Nonetheless, the significant presence of "ordinary people" in television and newspapers has done little to address problems of substantive political participation. Instead, citizens are largely represented as passive consumers to a world of politics which is incomprehensible, distant and outside their reach and influence. Rarely are they offered an opportunity to engage with political problems and offer their own solutions. They may be given a chance to respond to rail privatization and health services as *consumers,* commenting on the quality of train catering and the experience of having their blood pressure taken. However, they are not seen as having the necessary "expression entitlement" or the ability to comment on complex issues of policy which nevertheless affect them as citizens. This, instead, is left to professional experts—politicians, academics and business leaders.

The research nevertheless uncovered what is perhaps a surprising exception to this rule: In both the U.S. and UK contexts, where national citizens were all but barred from making any comment of political substance, citizens of foreign countries were widely seen commenting on political issues, whether it be French youth from poor Paris suburbs discussing their experience of racism and criticizing their government's failure to address it or Egyptians reflecting on their country's prodemocracy movement. Quite simply, foreigners were seen as legitimate sources for comment on national politics even if citizens of the media organization's own country were not. There may well be complex reasons for this practice—among them, in countries like the UK where broadcasters are governed by strict principles of political impartiality, any statement critical of the government would necessitate a response. Nevertheless, explicitly political statements by victims often feature prominently in disaster coverage and sometimes become a theme in news reports, as when coverage of the summer 2010 floods in Pakistan swiftly turned to a focus on local people's complaints about the government's failure to help those stranded without food, water or shelter in distant areas.

A more differentiated analysis suggests that disaster coverage makes distinctions between worthy and unworthy victims of disaster, based on prevailing power relations and geographical and cultural proximity. For example, poor African American victims of Hurricane Katrina were constructed as undeserving of assistance through accounts that emphasized rampant lawlessness (Garfield, 2007) in the wake of the disaster, creating a "disaster mythology" with detrimental consequences for the relief effort (Constable, 2008).The relationship between geographical and cultural proximity is central to Joye's (2009) study of disaster coverage in Flemish television news in 2006. The news stories he examined portrayed victims of flooding and landslides in Indonesia as passive and depersonalized pawns subject to the whims of nature (Joye, 2009). By contrast, the coverage constructed Australian and American victims of wildfires in ways that suggest they were deserving of compassion. They were represented as individuals with a name and a voice and the agency and skills to manage the disaster by themselves (Joye, 2009).

Thinkers have long recognized that compassion and solidarity are created through individualized storytelling which enables audiences to put themselves in someone else's shoes. Hannah Arendt (1968) thus suggested that storytelling about individual lives is able to connect to larger issues of the common good by penetrating "the meaning of what otherwise would remain an unbearable sequence of sheer happenings" (p. 104). As the narrator in Graham Greene's classic novel, *The Quiet American,* observed, "suffering is not increased by numbers, one body can contain all the suffering the world can feel" (cited in Benthall, 1993, p. 196).

Nevertheless, the politics of pity (Boltanski, 1999) generated out of stories about distant sufferers tends to be crafted out of a stock of geopoliticized discourses, often with a postcolonial tint. Rather than challenging global power relations, it might reinforce notions of the superiority of Western nations. Such a narrative does not invite compassion in the sense of genuine recognition of the fundamental rights of distant others (cf. Benhabib, 2004). Indeed, Rancière (2004) has been critical of claims around a cosmopolitan register, under which he fears that human rights become the rights of victims (cf. Fine, 2007, p. 80) or "of those unable to enact any rights or even any claim in their name, so that eventually their rights had to be upheld by others" (Rancière, 2004, p. 298). To Rancière, such representations work to justify humanitarian military intervention, in itself a project predicated on the naturalization of Western normative regimes and seen by critics as underwriting neo-colonial projects.

However, Chouliaraki (2006) has suggested that rather than assuming the best we can achieve is a politics of pity, we need to carefully distinguish among different types of disaster coverage and the forms of emotional response that they elicit.

To this end, she has proposed a typology of three different forms of news of suffering, what she calls "adventure," "emergency" and "ecstatic" news. Adventure news is "news of suffering without pity," characterized by brief, factual reports represented by a "void of agency," because of the absence of any human actors from the story. As a result, "there is no possibility of human contact between the other and the spectator." By contrast, emergency news is "news of suffering with pity," which represents victims as individuals who can be helped by the action of distant others. Finally, ecstatic news, exemplified by coverage after the September 11 attacks, brings the sufferers as close to the spectators as possible, opening up a space for identification (pp. 10–11). On the basis of her analysis of these three types of news, Chouliaraki concluded that news of suffering "reserves the spectators' capacity to connect for those who are like 'us' while blocking this same capacity for the largest majority of world sufferings—those experienced by distant 'others'" (p. 181). Of course, as later chapters discuss in more detail, the role of citizens in disaster is undergoing profound transformation. Among media scholars and practitioners, there is an increasing interest in ways to boost audience participation in media, through phenomena that are variously labeled as "citizen journalism," "produsage," "user-generated content" and "participatory journalism," among others (e.g., Bowman & Willis, 2003; Deuze, Bruns, & Neuberger, 2007; Gillmor, 2004). These changes are often described as radically disrupting the relationship between journalists and their audiences and challenging established hierarchies of voice and access. They have come to the fore in debates and news coverage in the light of the role of social media technologies like Twitter, YouTube and Facebook in calling attention to and enabling the democratic revolutions sweeping the Arab world in the spring of 2011. Nevertheless, in a global setting, the use of audience material—even if it democratizes aspects of news production—remains firmly embedded within existing logics and hierarchies of news production, as we discuss in more detail in Chapter 9. It continues to rely heavily on access to technology and forms of cultural capital which are unevenly distributed, thus frequently reflecting, rather than challenging, prevailing power relations.

Conclusion

This chapter has focused on understanding the journalistic production processes involved in disaster coverage and how they contribute to the construction of particular forms of citizenship. We have suggested that the mediated and mediatized nature of disaster reporting profoundly structures understandings of and engagements with the world. We only know of earthquakes in Japan, Haiti, the United

States and China, cyclones in Burma and the Asian tsunami because media organizations have invested resources in covering them. Conversely, events which do not achieve such coverage are rendered invisible and impossible for citizens to respond to and act upon.

The chapter has argued that the forms of coverage and, consequently, citizenship enabled by dominant media practices are overwhelmingly grounded in the nation state and its place within global geopolitical power relations. The geopolitics of disaster coverage has an impact on everything from the representation of victims to the deployment of organizational resources, and this means that geographically and culturally "distant sufferers" are accorded much less attention than more proximate ones and are consequently represented as unworthy victims. By contrast, the personalized and therefore worthier sufferers closer to home hail more cosmopolitan forms of citizenship among spectators who are able to act upon the suffering.

4

Emotional Discourses in Disaster News

DISASTERS BRING A FLOOD OF POWERFUL EMOTIONS TO THE PUBLIC SPHERE. A substantial amount of media coverage is devoted today to the emotions of those who have suffered or otherwise been affected by the disaster. It is through media representations that we bear witness to the shock, grief, fear and anger of the victims of disasters. It is also through processes of media representations that individual experiences and emotions become collective and political and gain larger meaning. Emotional expressions are at the heart of both media and humanitarian representations aiming to capture our attention and encourage people to engage. Obviously, not all disaster stories exhibit the same degree or kind of emotional storytelling. As we discussed in earlier chapters, alongside disaster reporting that contains highly emotional narratives of suffering and public response, there is disaster news that involves little or no emotion (see Chouliaraki, 2006).

In this chapter, we propose that emotions embedded in disaster narratives have a significant role in disaster communications, given that emotional discourses are integral to processes of assigning meaning, that is, to shaping how we understand and respond to specific disasters. Abu-Lughod and Lutz (2009) defined emotional discourses as "a form of social action that creates effects in the world, effects that are read in a culturally informed way by the audience for emotion talk" (p. 107). Thus, it is argued here that we must take the media content that seems to have "some affective content or effect" (Abu-Lughod & Lutz, 2009, p. 106) into account in order to fully analyze how disasters are defined, responded to and politically aligned. If we accept the idea that teaching how to feel and act "right"—a "sentimental education"—falls under the remit of journalistic work, alongside the traditional tasks of passing warnings and keeping people informed in the wake of a disaster, we need to consider how the emotive and informative are combined in mediated representations and how emotional coverage, including journalists' own emotions that increasingly, it seems, surface in the reporting, may shape the public's response to a given disaster.

Focusing on emotion is not a diversion from a politically engaged analysis but, as we see it, indispensable for understanding how media constitute disasters and how some disasters have the ability to mobilize global solidarity. It is not our intention to argue for the superiority of the emotions over reason but to discuss their essential role in making disasters mean and motivating ethical action. In our discussion we emphasize social and political aspects of emotions, which means that emotions are not seen solely as natural individual experiences but also as historically situated social constructs and practices that are taught and enacted through cultural narratives (e.g., Abu-Lughod & Lutz, 2009; Harding & Pribram, 2009). Such a sociopolitical understanding assumes that emotions work to establish and reinforce social relationships as well as to challenge existing power relations. If we treat emotions as discursive practices that reflect and reproduce society and culture, then the analysis of emotion goes beyond psychological perspectives (how those directly affected by disasters, journalists covering disasters, or members of the audience feel) to ask questions about how emotional discourses in the news media shape local and global audiences' engagement with distant suffering and how they are implicated in maintaining or challenging social hierarchies and power relationships.

We argue that emotions matter in disaster politics, first, because emotional discourses are motivating and moving. Briefly put, without emotion there is no commitment and ethical action. As the 18th-century moral sentiment theorist David Hume (1757) argued, "[R]eason, being cool and disengaged, is no motive to action." In recent analyses of the emotional dimensions of social and political life, emotions have been acknowledged as an important mobilizing force for political action and, therefore, perceived as potential drivers of social and political change (e.g., Barbalet, 1998, 2002; Goodwin & Jasper, 2003; Goodwin, Jasper, & Polletta, 2001; Gould, 2001). In terms of disaster reporting, the central moral question is whether mediated representations play a role in the political mobilization of citizens. In other words, do they encourage people to empathize with disaster victims and teach them, as Carrie Rentschler (2004, p. 300) wrote, "how to transform feeling into action?" Second, emotions are political given that they move people to seek attachments to some people and turn away from others. As scholars from various disciplines have argued, shared emotions toward "us" and "others" are an essential part of the constitution of local, national and global communities (e.g., Ahmed, 2004; Burkitt, 2009; Jasper, 1998). In disaster reporting, the pertinent question is how the audience is invited to feel for, and act upon, the suffering of people "who are not like 'us'" or how cosmopolitan sensibilities might be mobilized by news reports (Chouliaraki, 2006, p. 196).

Our view of emotions as social practices that situate us in relation to others also proposes that emotions are not irrational forces beyond our control, as they have been traditionally perceived. It is well established in contemporary research that emotion and reason are not in an oppositional relation but are mutually constitutive, emotions being neither more basic nor subversive to reason. Instead, as Alison Jaggar (2009, p. 64) asserted, they "reflect an aspect of human knowing inseparable from the other aspects." A third aspect of the social and political nature of emotions concerns their relationship with moral judgments: As proponents of cognitive theories of emotion argue, emotions are part of ethical deliberations involving evaluations about their object, such as what is right or wrong, fair or unfair, worthy or unworthy, and important or unimportant (Nussbaum, 2001). For example, feeling anger over the slow pace of disaster assistance may involve the evaluation that humanitarian aid agencies have failed to respond adequately. In the process of making this evaluation, people's emotional reactions and judgments are influenced by media frames and narratives. That is to say, feeling angry may be induced by news coverage that, for example, emphasizes aid organizations' distorted humanitarian principles or the failure of coordination and delivery (cf. Kim & Cameron, 2011; Nabi, 2003).

In this chapter, we first address the complex relationship between journalism and emotion. We also discuss how the extreme conditions presented by disasters affect the role of a journalist as an "impartial" professional witness and mediator who turns emotional testimonies of eyewitnesses into discourse (Ashuri & Pinchevski, 2009, p. 143). We then discuss the ways in which emotions are embedded in disaster narratives and how individual emotions enter the social and political realm through media representation. We also argue that emotional discourses in disaster narratives fuel and shape the emotions of local and global audiences. Subsequently, we examine the formation of feeling communities in and through the media in the aftermath of disaster, paying special attention to "ordinary people" as increasingly important news characters, sources and "participants" in disaster reporting.

Journalism and Emotional Expression

Disaster reporting is one of the few legitimate—though far from uncomplicated—places for emotional expression in news journalism. In disaster news, emotional content and address are justified by the emotionally charged topic as seen in the comment of a seasoned Finnish broadcast journalist: "I think it is odd to raise emotions in news reporting, apart from these tsunamis and such global disasters, where it comes naturally. Crying does not really belong in news" (cited in Pantti, 2010, p. 175). It is obvious, however, that journalists routinely assess stories for their dramatic and emotional potential, and hence, for their appeal to the audience. Yet, while expressions of emotion are the very epitome of the appeal of disaster coverage, they remain contentious, as focusing on individual and collective emotions also routinely raises critical questions around the ethics and quality of disaster reporting, typically in regard to the treatment of victims and disaster-affected persons or using emotive reporting to manipulate audience reactions.

While journalistic accounts widely accept that news needs emotion to catch and keep the attention of its audience, the role of emotions has received surprisingly little attention in journalism scholarship. Put differently, the focus of research has been highly selective, as the lion's share of emotion research within the field of journalism has focused on individual emotional effects, particularly on emotional responses to "bad news." Similarly, social science disaster research has discussed emotions mostly from the point of view of the individual since the early research in the 1950s. The rationale behind these studies has been to identify the best techniques for controlling fear and preventing panic and resentment (Quarantelli, 1989). In journalism studies, emotions have been typically discussed in terms of their deviance from the normative ideals of journalistic objectivity and impartiality. Accordingly, the "positive" connection between emotion and the public role of journalism, such as emotion's relationship with public engagement, has received significantly less attention (as an exception, see Costera Meijer, 2001; MacDonald, 2000). Journalism scholars have used the term "emotional" pejoratively so that it translates as "nonobjective," "entertaining," "sensational" or "manipulative." This is seen, for example, in how the emotional style of reporting has been used to make a distinction between quality journalism and tabloid journalism (e.g., Costera Meijer, 2001; Gripsrud, 1992; Harrington, 2008). While quality journalism is traditionally conceived as being characterized by abstract, rationalist analysis of social and political structures, tabloid news discourse has been described as antirationalist or sensational in that it puts emphasis on individual feelings and experiences (Connell, 1998).

In the context of disaster reporting, a common concern has been that emotion prospers at the expense of the "why journalism" that informs about what has happened and why and critically scrutinizes the disaster response, instead of focusing on the emotional reactions of the people caught up in tragedy, or of audiences, or—what is worse—of journalists themselves (see Kayser-Bril, 2008; Ward, 2010). Bob Franklin (1997), for example, argued that while broadsheet journalism aims to explain, tabloid journalism focuses on evoking emotional responses. Franklin takes an example from disaster news: He writes that the purpose of tabloid journalism is "less to inform than to elicit sympathy—a collective 'Oh how dreadful'—from the readership" (p. 8). This passage captures, beyond the concern of serious news becoming converted into entertainment, on the one hand, a long modern tradition that views emotion as compromising rationality and, on the other hand, an attitude that remains prevalent in both journalism research and practice, which considers that the purpose of journalism is *not* to evoke emotional responses from an audience.

In recent ethical assessments of disaster coverage, however, reason and emotion are increasingly described as intrinsically linked, even if the suspicion toward journalists' conscious elicitation of emotion remains. For example, Stephen Ward, in commenting on the Haitian earthquake coverage, argued that in disaster reporting, emotional and objective dimensions of journalism should converge into what he called a "humanistic journalism":

> A journalism of disasters is not a journalism of Olympian detachment. It is not a journalism fixated on stimulating the emotions of audiences. It is a humanistic journalism that combines reason and emotion. Humanistic journalists bring empathy to bear on the victims of tragedy—an empathy informed by facts and critical analysis. (2010)

Furthermore, journalists themselves seem to assess the role of emotion in news reporting and its relationship with "good journalism" in less antagonistic terms. A study of broadcast journalists' notions of the emotionality of news showed that while journalists in both commercial and public service news were critical toward "emotional news," and these views partly arose from traditional reason-emotion and information-entertainment binaries, they nevertheless believed that emotion does not challenge ideals of objectivity and truth-telling (Pantti, 2010). On the one hand, presenting and interpreting individual and collective emotions was seen as a part of journalism's aim to reveal reality, as "facts" without which the whole truth is not told; on the other hand, the main purpose of emotional storytelling was perceived as stimulating interest and facilitating the understanding of news. Emotion, then, was often described as a narrative technique that helps deliver information, provoke thoughts and capture attention. For example, emotional images of suffering people were seen as the most effective way of capturing viewers' attention and revealing the "truth" without any explanations or statistics. Journalists' views on the uses of emotion resonated with findings of studies that have examined viewers' limited capacity to process information in television messages. These studies have shown that emotion guides attention in such a way that more information is processed for emotional messages than for "calm" messages and that, furthermore, emotion may improve viewers' memory recall and comprehension (e.g., Lang, Dhillon, & Dong, 1995).

When it comes to the relationship between emotional storytelling and "good" journalism, the view shared by journalists was that whether or not emotion presents a danger to journalistic quality depends entirely on *how emotion is used and for what reasons.* The "right" and "wrong" uses were poignantly summed up by an investigative reporter working for the Dutch public service news NOS: "It all comes down to the question of whether journalism uses emotions to make the story more attractive or uses emotions to elicit emotions and get better ratings" (cited in Pantti, 2010, p. 177). This statement, however, also illustrates conceptual tensions in the professional discourse on emotion, especially regarding intentions and outcomes; journalists should produce "emotionally attractive" news stories to draw attention and retain interest but not in order to *consciously* elicit a particular emotional effect or seek commercial profit. Echoing journalism's traditional claim to be only the messenger of reality, the primary role the journalists ascribed to themselves was that of an uninvolved witness: The task of a journalist is to "register" emotions, not to actively manage public emotions, as Barry Richards has proposed in his book, *Emotional Governance* (2007). There were, however, obvious contradictions within this professional self-image. For example, journalists foregrounded the positive "impact" of emotional narratives beyond capturing the viewers' attention, such as raising awareness and compassion for the victims of disasters. Furthermore, journalists' discussion of appropriate and inappropriate uses of emotion showed that the "management" of public emotion is very much part of reporting on topics such as disasters, accidents and war, even if it is not labelled as such. This is seen, for instance, in how journalists highlighted the difference between international news and national news regarding the direction (positive/negative) and intensity (weak/strong) of emotion. Thus, while reporting on distant disasters is characterized by more negative and "hotter" (stronger) emotions, stories of disasters which we consider our own are bound to include more positive and calmer emotions because, as the editor-in-chief of Dutch public service news stated, they "simply concern people who themselves watch television" (cited in Pantti, 2010, p. 179). The different standards for emotional expression in international news and national news involve not only the more graphic depiction of horror for foreign disasters but also the more distressing, "raw" emotional disclosure of the distant victims and their families.

Emotional Journalists: Engaged Reporting and Feeling Rules

> Tradition says emotions can distort and color stories so they are no longer factual and accurate. The human journalist is asked to subdue his or her own perception of an object, even while he or she is asked to describe that object. Doesn't a person describe on the very basis of perception in the first place?...Journalists are, in fact, human, and they do...in fact experience emotions like everyone else. (Willis, 2003, pp. 14–15)

The fact that journalists experience emotions like everyone else is not surprising, and that they experience emotions in the extreme situations presented by disasters is even less so. However, as Jim Willis writes in his book, *The Human Journalist* (2003), the reporting of journalists has traditionally been expected to be grounded in facts and their presentation assumed to be rational or "cool" in a sense that journalists separate themselves from their emotions when covering stories (Schudson, 2001). While journalists see emotion as a legitimate part of disaster narratives—expressed for example by the testimonies of victims and eye witnesses or embedded in emotionally engaging images and sounds—they tend to share the view that journalists should not bring their own emotional responses into their reporting (Pantti, 2010). The normative distinction between giving a channel to emotions of people affected by a tragedy and making journalists' own feelings visible is drawn up clearly in BBC correspondent David Loyn's (2003) plea for objective journalism: "This [objective] approach is not dispassionate. It can be hugely passionate, requiring emotional engagement and human imagination. But it is not about my passion, how I feel. The viewer or listener does not want to know how I feel, but how people feel on the ground." The question we are dealing with here is what journalists have done—or what should they do—with the emotions they feel when reporting on disasters. What political contributions journalists might make when bearing witness to their own experiences rather than providing accounts of the suffering they have observed. How might that fuel audience engagement?

First, however, we should remind ourselves that news has always been emotional. It is not unusual to hear claims that news reporting used to be less emotional and more informative: "Unlike the past, contemporary news reports are swamped in emotion as if reporting and analysing feelings are the reporter's chief purpose"

(Mayes, 2000, p. 32). While different emotional discourses, vocabularies and "feeling rules" prevailed in the past, reporting on disasters has never been void of emotional narratives. Michael Barton's (1998) study on emotional discourses of *The New York Times* between 1852 and 1956 showed that in the 1850s, the reporter's empathy with the victims of tragedy was totally unconcealed, and readers were usually provided with what he described as a thorough "emotional investigation and modelling" (p. 158). This means that when disasters and accidents were reported, readers were told how they should feel, how the people affected felt and also how the reporter felt. In the turn of the 20th century, *The New York Times* moved toward a more emotionally restrained style, as the newspaper established its "modern voice," characterized by more sobriety, discretion and factual content when reporting on major disasters (minor disasters were treated in a more sensational style), even if not without occasional backsliding to the coverage filled with vivid emotional testimonies and morbid details (Barton, 1998).

Of course, there are variations from country to country and from publisher to publisher when it comes to emotional style of reporting. In our study of the coverage of major accidents in British newspapers *Daily Mail* and *Times* from the late 1920s to the late 1990s, we discovered a cyclical change in the amount of emotion talk together with a change in textual features that convey emotionality, signalling changes in feeling rules. From the 1920s to the 1950s, emotionality of news articles mainly stemmed from evoking the full horror of the event, describing in a detailed manner the agony and pain of the victims and the state of the bodies. The reporter of the 1920s did not hide his empathy with the victims or his own emotions, and readers were also specifically told by the reporter that the event was "sad," "dreadful" or "heartrending." *Daily Mail* coverage of the Paisley cinema disaster in Scotland, in which 71 children died on the last day of 1929, provides one example of a reporter who did not leave the readers to guess at his own emotions: "Early this morning I joined the relatives in their sad visit to the chamber of the dead. May I be spared from another such ordeal" ("A silent town," January 2, 1930). In the 1960s, the reporting of national disasters changed into a more detached, matter-of-fact journalism that gave less visibility to personal emotions and private experiences. Expressions of journalists' own emotional states were largely absent until the late 1990s, when they appeared again but were this time usually confined to journalistic columns and editorials. As we will see, the emergence of the public as the principal object of emotional storytelling in the late 1990s and early 2000s marks the biggest change in disaster texts' emotionality (Pantti & Sumiala, 2009; Pantti & Wahl-Jorgensen, 2007; Wardle, 2007).

In recent years, the discussion about a shift toward a more emotional and engaged reporting, in particular when it comes to the emotional topics of disaster and war, has intensified. For instance, CNN International's managing director, Chris Cramer, suggested that the Asian tsunami changed how journalists report the news:

> What has been different about much of the reporting, particularly on TV, has been that the emotional attachment between reporter and victim has been obvious. Gone is the professional, some might say artificial, detachment....Now, for the first time, media professionals are starting to tell us how they feel about some stories. And it will probably make them better journalists. (cited in Lyall, 2005)

It seems that emotional involvement of the journalist has become more acceptable or even expected in certain situations. What has changed is that there are new media platforms and forms representing different emotional styles that facilitate professional and nonprofessional journalists' emotional involvement in their stories (such as journalists' blogs). Journalists, then, seem to find increased cultural license and media opportunity to reflect on their own emotional turmoil and the dilemmas of witnessing traumatic disasters. Ben Brown, BBC foreign correspondent, has reflected on his personal involvement in different disaster zones around the world and the moral qualms of not "crossing the line" of objective reporting. Under the headline, "Yes, I was callous....Remarkable confession of guilt and shame by war reporter Ben Brown," *Mail Online* published Brown's "confessional article," which concludes as follows: "It's an extraordinary and privileged life, and you see history unfold before your eyes. But you also pay a price" (Brown, 2009). And Jonathan Rugman, *Channel 4 News* correspondent, reflected the following in a blog entitled, "Shaken but Not Stirred: Confessions of a Haiti Reporter":

> Journalists can be caught unawares, their professionalism quite possibly enhanced by explosions of sheer empathy. Jon Snow cried after one interview in Haiti. I cried last year after making a film about the sexual abuse of children in Kenya, so clearly this hack can crack in some circumstances, and without empathy, how are we going to make your viewers care, which they must if they are to donate money, which of course makes our jobs that much more worthwhile? And without empathy, how are you going to rise to the verbal challenge of matching the harrowing pictures your terrific cameraman has shot? (cited in Sambrook, 2010, p. 30)

Personal accounts and emotional confessions such as these increasingly circulate in online and traditional media where journalists proclaim not only that they care about the stories they are reporting on, especially when embedded in the disaster zone and confronting firsthand conditions of devastation and human tragedy but also give vent to their personal emotions when doing so. At the same time, some journalists take the opportunity to question the traditional norms of impartial, detached and objective journalism as they seek to steer a path between engaging audiences and hoping to elicit compassionate responses from them on the one side and remaining faithful to industry-based and professional expectations of emotionally detached journalism on the other. Carmit Wiesslitz and Tamar Ashuri (2011) argued that online journalism has fostered the emergence of a new journalistic model which they call "the moral journalist"; unlike the "objective" journalist who is supposed to remain outside of events, the moral journalists (who are, however, typically nonprofessional) present their personal experiences as they witness the suffering and pain of others with the aim of changing the witnessed reality. Surrounding shifts in the wider culture, above all the increased valorization of emotional expression, as well as changing priorities in the media industries, possibly encourage this increased journalistic reflexivity in respect to journalists witnessing in the field of disaster reporting. According to Howard Tumber and Marina Prentoulis (2003, p. 227), there is a general call today for a more "human" reporting that departs from an old style "founded on a 'macho' attitude that prohibited any display of emotion or psychological anguish." The traditional macho attitude the authors refer to is evident, for instance, in a study of the Dunblane Primary School tragedy in Scotland in 1996 (Berrington & Jemphrey, 2003). The general feeling among the reporters interviewed for the study was that journalists need to have the right guts for the job, that is, be able to cope with the event without letting it affect them personally. The reason behind this attitude is that disclosing emotional vulnerability has traditionally been—or has been feared to be—a career-breaker (Beam & Spratt, 2009).

The changing attitudes toward emotional engagement are perhaps best displayed in the discussion about journalism as a traumatizing profession (see Rentschler, 2009). This discussion, which has flourished in the past 15 years in both the research literature and documentation of journalists' personal experiences of war and disaster scenes, owes much to the research on post-traumatic stress disorder (PTSD). The literature on PTSD tries to determine the effects of violent conflicts and natural disasters on the mental health of those affected. The Dart Center for Journalism & Trauma, founded in 1999, has been central in "humanizing" the culture of crisis reporting and raising awareness about the impact of the media coverage of trauma and tragedy both on news professionals and audiences (Willis, 2003, p. 147). Journalists have been required to manage their emotions as part of their job in order to produce a neutral or "cool" emotional state. Obviously this does not mean that journalists have been immune to the stress and shock they are exposed to at scenes of disaster. For instance, Odén, Ghersetti, and Wallin (2009) studied the response of major Swedish news organizations to the Asian tsunami disaster and concluded that the journalists sent to affected areas were subjected to tremendous mental and physical strain. Many of the reporters were unprepared for the horrors they encountered, and they found it particularly difficult to deal with the extremely distressed Swedish victims who approached them as fellow human beings rather than as professionals at work. As a television reporter explained, it can be a struggle to remain detached under intense pressure and fatigue:

> This was the worst experience I have ever had. When I've been reporting from war zones, I've been prepared for dying people, collapsing houses and desperate people. But this was something different; fellow countrymen, a tourist's paradise where mums, dads, children and friends go to enjoy the sun.... But you have to try to keep distance. In the beginning, when I felt fit, it was all right, but then it just got more and more difficult. (cited in Odén et al., 2009, p. 203)

With the growing awareness that reporters are subjected to traumatization when they are testifying to the suffering and violence, it has been easier to see journalists as vulnerable human beings who feel like the rest of us (Beam & Spratt, 2009; Clarke, 2005; Rentschler, 2009; Santos, 2009; Willis, 2003). However, as Carrie Rentschler (2009, pp. 177–178) pointed out, this trauma discourse "locates the problem of the intensified industrialization and commercialization of catastrophe news in the effects it has on individual workers and not in the structural conditions that make news work and the events being covered more trauma prone."

In the more "human" culture of reporting, becoming emotionally involved in a story is not automatically deemed unprofessional. As journalism educators Tom Rosenstiel and Bill Kovach (2005) noted in their comment on the surge of emotion among U.S. journalists who covered Hurricane Katrina in 2005, "Journalists are in essence our surrogate observers. It would have been odd, even distressing to most, if reporters had reacted like journalistic robots to the devastation in the Gulf Coast." Recently, we have also seen that particularly emotive disaster reports tend to receive a lot of attention and are often appreciated by both journalists and audiences. One much-discussed example where a journalist "crossed the line" from the position of the distant observer was BBC reporter Ben Brown's interview in Aceh in Indonesia with a woman who had lost her husband and four children in the 2004 Boxing Day tsunami. A neutral interview turned into an awkward act of comforting as Brown put his arm around the woman who was weeping and clinging to him. The appeal of this footage was described by a managing editor for CNN, Nick Wrenn: "The Ben Brown incident really brought it home to me—just the raw emotion of it. And it is an emotional experience, and I think sometimes you have to engage. To me that was one of the most powerful packages that I saw out of all the broadcasters" (cited in Dart Center for Journalism & Trauma, 2005).

However, this new legitimacy of journalists' own emotional expressions requires further attention. It is clear that the emotional display of journalists, even if deemed appropriate or necessary in certain circumstances, remains highly controversial. The increased cultural licence to display emotions is also unlikely to produce a general shift in attitudes of journalists toward traditional values of emotional detachment and objectivity. As we will see in Chapter 5, based on the accounts of the correspondents and journalists themselves, most journalists continue to subscribe to ideas of news objectivity in terms of impartiality and detachment while nonetheless conceding that not only do they care but they want their news reports to help audiences care as well. A recent example of the controversy surrounding "emotional journalists" comes from the coverage of the Haiti earthquake in January 2010: A YLE (Finland's national public service broadcasting company) reporter was seething with anger on the air when reporting from the Haiti airport because of the delay of aid and what she described as a "hellish" situation in Port-au-Prince. The report prompted extensive online discussion in Finland where the outburst was both applauded and criticized. Some wrote that breaking away from the routine disaster reporting with an unscripted expression of emotion had motivated them to donate money, while others questioned why a public broadcaster had sent an unprofessional "girl" to the disaster area.

There is no normative consensus over what emotions journalists ought to display, and when and how, but the issues around journalists' emotional expression that are typically discussed include the authenticity of emotional responses, the "right amount" of emotion for the situation, and the motivation behind the display of emotion (Pantti, 2010). All these aspects are present, for example, in the feeling rules for disaster coverage outlined by Rosenstiel and Kovach (2005) in the aftermath of Hurricane Katrina and by Ward (2010) in the aftermath of the Haiti earthquake. The first rule concerns situational appropriateness and authenticity. It means that "emotion ought to come at those moments when any other reaction would seem forced or out of place" (Rosenstiel & Kovach, 2005). In other words, emotional expression is legitimate when it is the only natural response, as opposed to being a journalistic gimmick used to appeal to audiences. The second rule concerns the proper "amount" and "duration" of emotion. One of Rosenstiel and Kovach's (2005) rules is that journalists should quickly compose themselves in order to focus on the how and why questions: "[O]nce journalists have reacted in a human way to what they've seen, they must compose themselves to sort out responsibility for how and why things happened. The search for answers requires all their scepticism, professionalism and intellectual independence." Finding the right balance between objective information and emotional accounts seems to be paramount. As Ward (2010) argued, journalism ethics requires that emotions are used proportionately in disaster coverage: "Emotional stories should not dominate the news coverage and overshadow critical analysis." The third rule concerns journalists' motivation and role, which, according to the ideals of objectivity, should not be in the center but "on the sidelines…of society's dramas" (Schudson, 1995, p. 52). As we have seen, journalists are criticized for making their own emotions the eye of the story rather than the emotions of the people who are actually suffering:

> Journalists should not use emotions to make themselves the center of the story and to engage in self-congratulation. In an era where the use of media is "all about me," disaster coverage needs to move in the opposite direction by focusing on the story, not the story tellers. (Ward, 2010)

What can be gained from employing a more emotionally oriented approach to reporting? First, some have pointed to the increased lack of trust in the news media and to the fact that the traditional, objective mode of journalism failed to prevent journalists from losing their audience. Furthermore, some have claimed that allowing journalists' feelings to fuel the reporting is necessary if news media are to reconnect with their audiences (Santos, 2009; Willis, 2003). Second, some have argued, following from a consideration of journalists' ethical obligation to manage audience encounters with traumatic events, that an emotionally engaged or sensitive approach renders journalists better able to manage and contain both their own and the audiences' feelings (Tumber & Prentoulis, 2003; Richards, 2007). Third, a more emotionally engaged approach has been seen to help generate a new moral imaginary when bearing witness to human suffering. It has been argued that journalists should learn to care more about their stories and "allow themselves to feel the anguish of what it's like to be a victim" in order to help people see injustice and suffering as something that must be addressed instead of offering them comfort or excitement (Santos, 2009, p. 39). As Willis (2003) claimed, "readers and viewers become more engaged in a story if they feel the reporter cares about the story, the people in it, and cares enough about the reader or viewer to write a story evidencing that care" (p. 129). The question that follows from this idea of making disasters matter to the public through emotionally engaged reporting is the same one we discussed before and continue discussing in later chapters: Why do some disasters bring out more emotionally engaged (and engaging) reporting than others?

Emotions in Disaster Narratives

In the literature on the witnessing of distant suffering, the crucial question has been how media reports on distant suffering can mobilize a moral response on the part of the media publics (see Boltanski, 1999; Chouliaraki, 2006; Silverstone, 2007; Tester, 2001). The basic conviction behind this body of work is that media *representations* in fact have the potential for shaping our engagement with the distant sufferer. Thus, there is a crucial link between representation and emotion/action, and respectively, between representation and indifference/nonaction. As our focus here is on the role of emotion in the meaning-making processes of disaster reporting, it is clear that we need to examine the ways in which emotions may manifest in news reports and how disaster narratives might be read as encouraging certain emotional responses and discouraging others. Overall, we may say that representations and narratives are our prime collective resources for examining and understanding emotions that are fundamentally "unknowable" and "unshareable" experience (see Bleiker & Hutchison, 2008). The emotions of people affected by a disaster become visible and "knowable" only in stories about emotion, and, as we stated in the introduction of this chapter, it is in and through the processes of representation that emotions translate into a larger political discourse that shapes the response to a disaster.

The emphasis on language and discourse in studying emotions opens up insights into how mediated emotions reconstitute social groups and reiterate cultural values, norms and meanings. The statements and condolences expressed by the national and foreign political elite in the aftermath of disasters and accidents are a stable narrative element in disaster reporting, that together with representations of different mourning rituals and symbols, from flying a flag at half mast to constructing public shrines, signals the significance of the event to the wider public and at the same time constructs a discourse of communality. We can take the Prince of Wales' response to the Paddington train crash in London in 1999 as a typical example of how expressions of emotions embody cultural values and produce social identities: "My heart goes out to all the relations of those who lost their lives in this dreadful crash. These terrible things bring out the best in people—I heard wonderful stories about passengers helping other passengers" (*Daily Mail,* "My sorrow, by Charles," October 7, 1999). The statement from Prince Charles articulates shock and sympathy and is explicitly oriented toward making sense of the "dreadful crash" and "terrible thing" by identifying the audience's moral character ("bring out the best in people"), and in so doing, enacting a sense of solidarity, heroism and national pride.

Representations of emotions are powerful means for imagining unified communities and offering normative scripts of how we should respond. However, disaster reporting articulates not only collective and ritualistic performance of grief and solidarity that create a discourse of communality. The rhetoric of unity through emotion is also used to amplify anger and rage against the power holders to constitute a discourse of crisis. In the following editorial about the Paddington crash, the public anger described by the reporter is explicitly linked to the criticism of government's inadequate efforts at addressing rail safety (see Chapter 8), and in the following report on the Haiti earthquake, violent anger is aimed at the government, the United States and international humanitarian relief organizations:

> Understandably, the first feelings of public shock over the Paddington rail tragedy are turning to seething anger. How could such a disaster happen, in the light of the seemingly very similar Southall crash two years ago? Why has there been such official reluctance to install the Automatic Train Protection system?...Why was no action taken, despite the fact that eight trains have passed it at red in the last six years?...Meanwhile worried commuters accuse those companies of putting profits before safety. Deputy Prime Minister John Prescott faces mounting criticism for doing so little to improve standards. And some Labour MPs are denouncing the Tories for privatising the railways and supposedly putting safety at risk. (*Daily Mail,* "Why Parliament must be recalled," October 7, 1999)

> Protests over the slow arrival of aid have flared in Haiti's capital Port-au-Prince as Prime Minister Jean-Max Bellerive put the earthquake death toll at more than 200,000...."If the police fire on us, we are going to set things ablaze," shouted one protester, raising a cement block above his head.
>
> Another 200 protesters marched to the US embassy, crying out for food and aid, while about 50 demonstrators gathered outside the police headquarters where the government of President Rene Preval is temporarily installed. "Down with Preval," demonstrators shouted at the President, who has spoken to the people only a few times since the disaster. (*The Age,* "Haitians angry over slow aid," February 5, 2010)

All narratives are intimately bound up with emotion in that they depict and stimulate emotions. But that is not all there is to emotions and narratives. Our argument is that emotional responses to disasters are guided by narrative scripts, that is to say that news reports lead us toward certain emotions and "suggest" the feelings that are suitable in a given situation. News media are not only reporting on emotions but also generating them and educating us about them and thus have an important role to play in the global emotional politics of disaster. Tony Walter, Jane Littlewood and Michael Pickering (1995) have employed the concept of "emotional invigilation" to describe the twofold role of the media in *mobilizing* and *managing* emotions related to death and suffering. One of their arguments is that with news reports covering death, certain affective dispositions are mobilized and certain emotional responses encouraged while others are discouraged. Of course, one cannot assume that audiences' emotional responses to a disaster event or identifications with disaster victims or other characters depicted in news stories are automatically overdetermined by the news texts, just as one cannot assume that depictions of emotion in disaster reporting accurately represent individual or collective emotions. Even if a disaster is defined as a tragedy that affects us all, it does not follow that audience members all feel the same—or anything at all. However, like films and other dramatic texts, news reporting is a form of storytelling which is *emotionally prefocused* to elicit certain kinds of responses (Carroll, 1997). The issue of how media representations shape our emotional responses can also be approached from the perspective of "feeling rules" and "expression rules" (Hochschild, 1979) that exist as part of our emotional culture and construct our emotional experiences (see Gordon, 1990). Sociologist Arlie Hochschild draws attention to the fact that we do not solely feel emotions but have clear ideas about their appropriateness and morality. Feeling rules and expression rules function as norms regarding appropriate affective responses in a given situation and when and how to display emotions (Hochschild, 1979).

It is important to note that the factualness of suffering in news texts, in comparison to the death and suffering in fiction, limits the emotional responses that news reports encourage. As we discuss in Chapter 6, disaster reporting is informed by the moral norm of compassion, and therefore we are, above all, invited to feel compassion for the sufferers (Chouliaraki, 2009, p. 521). We can begin the discussion about this difference by looking at Hume's (1757) account of the extraordinary pleasure the audience receives from painful emotions in "well-written" tragedies:

> It seems an unaccountable pleasure, which the spectators of a well-written tragedy receive from sorrow, terror, anxiety, and other passions, that are in themselves disagreeable and uneasy. The more they are touched and affected, the more are they delighted with the spectacle.... The whole art of the poet is employed, in rousing and supporting the compassion and indignation, the anxiety and resentment of his audience. They are pleased in proportion as they are afflicted, and never are so happy as when they employ tears, sobs, and cries, to give vent to their sorrow, and relieve their heart, swoln with the tenderest sympathy and compassion.

What Hume argues in his essay "Of Tragedy" is that the unpleasantness of experiencing negative emotions in response to the tragic events depicted in the play is overpowered by the great pleasure derived from the artistic manner of representation. However, a precondition for this pleasure of experiencing painful emotions such as grief, distress and terror is that they are aroused by fictional tragic incidents rather than real events. As Walter and his colleagues concluded, "News audiences are likely to experience not so much personal pleasure as vicarious pain on behalf of those suffering and/or anxiety that this could happen to them or to their children" (1995, p. 586). Of course, what the authors describe is the socially appropriate affective response. It is clear that the same narrative or image of suffering can make possible very different emotional responses from compassion and indignation to annoyance, disgust and indifference. We are, however, taught or socialized to feel compassion, and disaster news can be understood as a discursive resource for "feeling right," that is, for offering socially accepted and expected emotions.

Another important difference between the emotional responses elicited or encouraged by news texts and fictional texts is that fiction-based emotions do not call upon the audience to respond in terms of action. The modern conception of compassion informing disaster reporting involves the idea that we must in some way respond to the suffering we have come to know through the media (Ellis, 2000). One of the purposes of emotional discourse in disaster news is to mobilize public action, typically through donations to disaster relief funds but also through engaging in public debate, for example around the moral obligations we have, the range of action we can take or around the adequacy and efficiency of disaster response (Boltanski, 1999). Luc Boltanski (1999) sees that commitment (including speech on behalf of the sufferers) leads to a "moment of transformation from the state of being a receiver of information, that is to say, of being a spectator, observer or listener, into that of being an actor" (p. 31). However, the issue of the powerless spectator of fiction is central also for witnessing real suffering, since the distinction between reality and fiction may lose its relevance when the sufferer and the possibility of action are far away (Boltanski, 1999, p. 23; Moeller, 1999). There is a strand of studies that follows upon Baudrillard (1988) and concern the voyeuristic pleasure of witnessing suffering in a situation when disasters are conducted as media spectacles that, on the one hand, blur the distinction between reality and mediation/entertainment and, on the other hand, generate unbridgeable distance between the safe spectator and the suffering victim.

Considering that emotional discourses and scripts shape the emotional responses of the public, we need to examine how emotions are inscribed in the conventions of disaster narratives and how they are described and ascribed to individuals and collectives. First, the question about *what* emotions are available to be felt or spoken of in disaster reporting is important because particular emotions are associated with particular behaviours and meanings. So, we need to see that choices have been made regarding the representation and encouragement of particular emotions because different emotions help to make certain actions and political positions possible (see Loseke, 2009; Skitka, Bauman, & Mullen, 2006; Small, Lerner, & Fischhoff, 2006). In the context of 9/11, Linda Skitka and her colleagues (2006) examined how different emotional responses led to different political positions: "Anger but not fear predicted support for expanding the war beyond Afghanistan, and fear but not anger predicted support for deporting Arab Americans, Muslims, and first generation immigrants." (p. 375) It is, however, important to note the complexity of emotional politics and draw attention to the fact that the same emotion can serve a variety of political purposes. Anger is often classified as a "negative" emotion inspiring acts of violence, but like compassion it can invoke concerns with human suffering and injustice. Anger, as we discuss in more detail in Chapter 8, can play an important political role in seeking social justice and challenging power relations. Similarly, as Judith Butler argued in her book, *Precarious Life* (2004), grief and mourning can represent an impetus for people to act politically and rethink international politics in a more ethical manner—to imagine a world "in which an inevitable interdependency becomes acknowledged as the basis for a global political community"—but grief also works to justify violence and define who counts as human (who is "mournable") and who is excluded from humanity (pp. xii–xiii).

Disaster news represents and channels a wide range of emotions. As we have argued elsewhere (Pantti & Wahl-Jorgensen, 2007), high-profile disasters are usually inscribed with four main emotion discourses. The coverage of a disaster typically opens with a discourse of horror that depicts the dreadful consequences of the tragedy: death, destruction and debris. The description of horror in disaster news, with shocking images of pain, works to generate a sense of "moral shock" among audience members that could lead to a willingness to act (Jasper, 1997, p. 129). The discourse of horror is accompanied by the discourse of grief, which focuses on the bereaved families and communities. Journalistic discourse of grief finds its expression not only in the narratives of personal loss but increasingly also in collective mourning rituals and commemoration, which, in the Durkheimian sense, work to reinforce solidarity. Such accounts, in turn, give rise to the discourse of compassion, which focuses on the need or desire to alleviate suffering and routinely emphasizes the efforts of heroic individuals and aid organizations. Finally, the discourse of anger focuses on questions of blame and responsibility by telling stories of the anger of the afflicted or the broader public outrage in response to the disaster.

Emotions are attached to historically and culturally changing values and moral norms, which shape the way that particular emotions are represented, experienced and managed. We can illustrate this historical and cultural specificity by taking the discourse of horror as an example. Western mainstream media representations of disasters, accidents and wars have been claimed to be increasingly sanitized (Taylor, 1998). There seems to be a move toward a more sensitive depiction of horror and anguish in disaster news, at least when it comes to covering disasters that can be considered "ours." Our study of the coverage of British national disasters shows that the language used when reporting death used to be more vivid and shocking (Pantti & Wahl-Jorgensen, 2007). To accord decency to the victims and respect to the bereaved, journalists covering national disasters today refrain from displaying graphic details and using sensational descriptions such as "rotting flesh," found, for instance, in the *Daily Mail* story of the London Tube disaster in 1975: "They kept spraying disinfectant and then deodorant but the piped air from the pavement kept turning things around and rotting flesh was the smell that made the workers retch" ("Inside the Tunnel of Death," March 3, 1975). In the coverage of the Paddington train crash in 1999, the depiction was less focused on bodies and horrific details than in reports from earlier decades. Journalists were more likely to refer to the death through symbolic representations, such as writing about commuters' cars with frozen windows abandoned in the railway station's car park and cell phones ringing unanswered in the train wreck.

We have argued before that news media coverage of disasters that occur far from home has more intense depictions of emotions. However, it is not necessarily the case that the further away the suffering, the more graphically it will be covered. Folker Hanusch's studies (2008, 2010) showed that newspapers cover death in quite different ways around the world. His study of how German and Australian newspapers cover death in international news coverage showed that there are national differences in the acceptability of sensationalist language (Hanusch, 2008). For example, Hanusch discussed responses to the headline "Hundreds of rotting bodies in Haiti city" of a story about the dead that remained unburied for days after Hurricane Jeanne hit Haiti in late September 2004 (published in *The Australian* on September 24, 2004). While some (German) journalists were upset about the use of the word "rotting," the headline was generally accepted because it accurately reflected what happened, and the word was not used merely as a sensationalist device. Another study examined newspaper coverage of the 2010 Haiti earthquake in 24 quality newspapers from 12 countries in order to compare how the dead are visualized in disasters. The findings show that the coverage is affected by the media system and religious standards. Newspapers in Catholic countries that represent the Polarised Pluralist system (Brazil, Italy, Mexico, Spain) showed visible bodies or contorted bodies more often than newspapers in Protestant countries and/or those representing the Democratic Corporatist model (Belgium, Germany, Norway, Switzerland), which preferred to show bodies covered with sheets or from a distance or to show only body parts. The value of horror in the representations of suffering has prompted a long debate. While graphic images and graphic narratives are commonly understood as leading to "compassion fatigue" (Moeller, 1999), some have seen the display of horrific imagery as a tool to overcome news fatigue and motivate action against injustices (Campbell, 2004; Taylor, 1998).

At this cultural moment, however, the discourse of horror is going through a change, as new technology (digital photography, the Internet, etc.) is reshaping the ways that death and suffering are represented. In recent years, the explosion of citizen photography and video has become a crucial resource for disaster reporting, and this type of imagery increasingly serves to shape our encounter with and knowledge of traumatic events both distant and close. Importantly, the citizen material goes beyond traditional journalistic boundaries, which makes the visual coverage of death and suffering potentially more confrontational and graphic. The question that needs to be addressed is what role citizen eyewitness photographs and videos will play, on the one hand, in the newsworthiness of a disaster and, on the other hand, in the development of audience emotional engagement with distant sufferers.

Second, besides looking at which emotions are depicted and reinforced in disaster reporting, we need to ask *whose* emotions are considered worth acknowledging—whose emotions become visible and prominent in public discourse? We can identify different "emotional actors" in disaster coverage. Traditionally, the most important of these are the victims and eyewitnesses. However, the degree to which victims are present in coverage, and the reasons for including them, have changed over the years (Pantti & Wahl-Jorgensen, 2007). Until the 1990s, the main purpose of interviewing victims was to provide information about the disaster and rescue. Journalists were eager to depict what the victims had seen or heard and therefore collected multiple eyewitness accounts in order to give a more objective description of the event. During the past few decades the number of witnesses has decreased as the media has focused on individual tragedies and rescue stories. At the same time, the primary function of victim and eyewitness interviews has changed from providing information about what happened to telling what people feel.

The emotions of different actors matter in terms of their legitimacy and visibility in the mediated public sphere. They can serve as a means through which the suffering groups can visit the symbolic space of media and speak and be heard (see Silverstone, 2007, pp. 139–142). The way disaster reports describe the emotions of victims or other news characters has an effect on how we see them as political and moral actors. It has been argued in the literature on media audiences' relationship to the human suffering that presenting sufferers as active agents (as opposed to "passive" victims) is crucial to the development of compassion (Chouliaraki, 2006, pp. 88–90; Tester, 2001). Given that emotions cannot be separated from action and agency, presenting the victims with humanizing agency includes allowing them to express a full range of emotions generated by the disaster.

It is also important to note that the public display of emotion by different groups does not necessarily carry the same symbolic value, emotional stereotypes related to gender and social status being prominent examples. For instance, certain emotions have been considered more appropriate for, or typical of, women or men: for women, these have included "powerless" emotions such as grief and fear and, for men, "powerful" emotions such as anger (Lupton, 1998; Woodward, 2004). Cultural expectations regarding expressions of emotion, together with cultural constructions of "worthy" and "unworthy" victims, are seen in Birgitta Höijer's (2004) study of Swedish and Norwegian audience reaction to media coverage of human suffering in the Kosovo war: While images of crying children and elderly people left a long-term impact on the emotions and memories of the interviewees, a crying middle-aged man in a refugee camp who begged to be brought to safety prompted resentment among them. Clearly, there has been a disparity in the witnessing of suffering as the emotional testimonies often come from the representatives of Western NGOs. The boundary between the "silent sufferer" and the "speaking Westerner" is not necessarily disappearing with the new media technologies and online platforms that have allowed for the proliferation of ordinary voices (Chouliaraki, 2010). While new audiovisual recording technologies have empowered people to make their voices heard, we should keep in mind that the "digital divide" still exists, and access to this technology is not democratically distributed. For example, Chouliaraki's (2010) study on BBC convergent news of the Haiti earthquake shows that web-streaming invites the contributions of the affected as the main source of news, but the vast majority of testimonial messages come from NGOs and other Westerners indirectly affected by the disasters.

Emotional Publics: Participation and Public Therapy

In James Carey's (1989) view, communication is a ritual process that articulates shared meanings and "draws persons together in fellowship and commonality" (p. 18). This social function of news is especially evident in the coverage of disasters—both distant and close—as the disaster narrative commonly includes journalists' attempts to make horrific events understandable and offer consolation by employing established cultural scripts (such as the aforementioned idea that "disaster[s] bring the best out in people"), using certain kinds of symbols and focusing on the emotional display and ritual activities of ordinary people in the aftermath of disaster. We discuss the formation of national and global feeling communities from the point of view of the moral obligation to feel compassion and act toward relieving suffering in Chapter 7. Here we touch on the role of disaster news in forming communities in the aftermath of disasters from the point of view of media witnessing. In particular, we look at the proliferation of ordinary voices in disaster reporting and its role in the formation of ad hoc affective communities.

Disaster news is a source of anxiety in that it constantly reminds us of our shared vulnerability, but as several scholars have noted, it is also a site for the management of anxiety: It comforts us and helps us to "work through" what we have witnessed through news coverage (Ellis, 2000; Walter, 2006). In a media-saturated global era, as Paul Frosh and Amit Pinchevski stated, crises and disasters are no longer interruptions into our lives; they have become a routine, experiential ground of it. According to them, media witnessing incorporates audiences into "a perpetual *crisis-readiness*" and helps to shape a global sentiment of shared human vulnerability and impending threats (Frosh & Pinchevski, 2009, p. 296). This means that the trauma discourse we discussed earlier in the context of the journalism profession is extended to audiences: To witness crises through the media is to be traumatized. This feeling among viewers that they are traumatized by what they see, read or hear is about "mediatized *enabling* of crisis-emotions among those who were not physically present at the event but nevertheless feel themselves affected by it ('this could very well happen to us')" (Frosh & Pinchevski, 2009, p. 301).

We can approach this extension of trauma discourse and proliferation of crisis-emotions among global publics in terms of the prominence of the first-person "emotion talk" in the emotional culture we are living now (e.g., Lupton, 1998). At this cultural moment, we are increasingly displaying and reflecting on our emotions in the media, individual emotional experience having become fundamental to self-representation via the therapeutic narrative (Illouz, 2007). In his analysis of "first person" media, John Dovey (2000) offered the conclusion that the proliferation of publicly mediated, individual experiences and emotions operates as a new regime of truth and guarantor of authenticity (pp. 23, 25). The appearance of directly uninvolved people as witnesses in the media is a relatively new element in disaster news. Since the 1990s, national and global disasters have given rise to ritual expressions of public emotion such as laying flowers, signing books of condolence, or participating in silences and civic or religious ceremonies (Pantti & Wahl-Jorgensen, 2007). The closing decade of the 20th century saw an increase in ordinary people invited to speak for themselves and express their feelings with their own voices in disaster reporting. The intense emotions generated by disastrous events, in particular "good," comforting feelings such as the "waves of compassion," national solidarity or patriotic pride that follow disasters, have become newsworthy in their own right (see Kitch, 2003). As Carolyn Kitch (2009) wrote, in the context of U.S. journalism, tragedies are now produced in news "as a story of the living who grieve, a story about public ritual. The focus will be on mourners who erect spontaneous shrines, who stand outdoors holding candles or American flags, and who very publicly pay tribute to the dead" (p. 29).

The flourishing of the emotional expressions of "ordinary people" in disaster coverage justifies claims of a new metalevel in the emotional storytelling of disaster news. At the present time, there are clearly increased and emerging opportunities for emotional expression and participation in both off-line and online activities following disasters. Like social networking sites, online news sites create a forum for people who have in some way been touched by a disaster and who need to publicly talk about their feelings and/or acknowledge the pain of others. Shane Bowman and Chris Willis (2003) wrote the following in their citizen journalism manifesto, *We Media:*

> The response on the Internet gave rise to a new proliferation of 'do-it-yourself journalism.' Everything from eyewitness accounts and photo galleries to commentary and personal storytelling emerged to help people collectively grasp the confusion, anger and loss felt in the wake of the tragedy. (pp. 7–8)

Frosh and Pinchevski (2009) argued that we need to understand crisis not only as something that media audiences "see" but as something "they are increasingly socialized to create," as the ultimate witnesses of events but also as the most important producers of mediated testimonies (p. 300). With new communication technologies enabling new models of behaviour, members of the public are playing today an increasingly active and manifold role in disaster response and recovery (e.g., Hughes, Palen, Sutton, Liu, & Vieweg, 2008). The argument we want to pursue here is that ordinary people are now increasingly appearing as models for the appropriate emotional response during times of disaster, whether it comes in the shape of *mourning,* such as creating memorial sites (virtual or real); *supporting,* for example, showing support for disaster victims or official response personnel; or *helping* victims, for example, by providing information about sources of relief or making humanitarian appeals in the social media platforms, which previously was the domain of the aid agencies.

That people are speaking for themselves, rather than being represented by media professionals, invokes the idea of the democratic potential of media participation as well as the therapeutic aspects of self-expression (e.g., Thumim,

2010).The case of the ABC News online coverage of the Victoria bushfire emergency in 2009 offers an example of the informational and emotive elements of citizen participation in disaster reporting today. It also highlights the role of public emotion in the building of affective communities and the containment of anxiety caused by a disaster. The eyewitness accounts of ordinary citizens played a key role in the coverage, because the disaster area was difficult for journalists to access. During the first days the audience material consisted primarily of eyewitness information such as video footage, photos and phone calls that was used to provide information about the firefighting efforts, the physical effects of the fires on properties and landscapes, and account for the experiences (like people escaping from the fires). The following is a call from an ABC Local Radio listener, posted in the "ABC Bushfire Community" site which asks you to share "your experiences by text, photos, audio or video":

> Peter: The whole of Kinglake is ablaze mate. I live a couple of k's out of town and I heard explosions so I just went to the end of my road to see what was going on...and there were fires everywhere and I just bolted back. . . .
>
> Jon Faine (head 774 ABC radio presenter): What can you actually see now?
>
> Peter: Flames everywhere. Trees exploding, gas tanks exploding. It's really very, very serious. ("Kinglake is ablaze mate," February 7, 2009)

In this example the focus is on the question, "What can you see?" and accordingly, the eyewitness account is a source of information that makes a contribution to the ABC reporting of breaking news, not least by adding a sense of authenticity and immediacy based on the caller's firsthand experience of the fires. This is an example of "participatory journalism," in which the citizen participation takes place within the framework and control of professional journalism. As such, it fundamentally replicates traditional newsroom hierarchies: The professional journalist manages the media report in which the ordinary person appears (Couldry, 2003). Journalists use citizen-created audiovisual content strategically to tell the stories in an attractive way and to add a sense of "reality" (Pantti & Bakker, 2009; Williams, Wahl-Jorgensen, & Wardle, 2011). Audience research findings showed that members of the audience, downplaying the importance of the traditional detached mode of journalism, appreciate the perceived authenticity of such material and, above all, the emotional engagement invited by seeing events through the eyes of "ordinary people" (Williams et al., 2011). Andrew Williams and his colleagues (2011), however, concluded that journalists relegate the role of citizen journalist to that of "a provider of supplementary emotional and emotive rather than rational and political material" in order to protect their privileged status as producers of knowledge. The distinction between "emotional" and "political" is problematic, as we have discussed in this chapter. For instance, the mobile phone videos showing the death of Neda Agha-Soltan in June 2009 by an Iranian government sniper illustrate how dramatic audiovisual amateur material can frame the political discussion and produce emotional *and* political communities of solidarity and protest on a global scale. Amateur videos of the death of Neda Agha-Soltan also illustrate how the ubiquitous audiovisual technologies blur the distinction between firsthand and secondary witnessing: We are interpellated to conceive ourselves as the ultimate witnesses of events rather than members of a detached and distant audience (Frosh & Pinchevski, 2009).

The ABC Bushfire Community also is an example of media space where there is no hierarchy of discourse and where access is unlimited. By comparison to traditional broadcasting, on the ABC site there are no participatory constraints placed on the ordinary person. How it works as a space for community formation is evident in the material that began to appear after the acute phase of the disaster. Gradually the form and style of the messages in the ABC Bushfire Community changed from firsthand eyewitness accounts into more emotionally reflective contributions. There was an increasing number of user-composed poems and song lyrics as well as condolences and prayers. The greater part of the contributions posted a couple of weeks after the fires broke out across South Australia was made up of reactions to and reflections on the earlier messages, many expressing appreciation for shared experiences, solidarity and reassurance, like the following:

> Like everyone, I have been deeply affected by the sheer enormity of the recent bush fire tragedy and felt I wanted to express my thoughts as the days pass. Nothing will remove the incredible loss and sadness for those involved, but in seeing the wonderful, generous response to those in their greatest need I take some comfort in knowing there are so many people who have opened their hearts to help their fellow human beings, in their deep stress and sadness. Surely we are witnessing "The Hand of God." ("Hearts full of sorrow," February 19, 2009)

Unsurprisingly, the bushfire disaster intensified national feeling, and patriotism was prominent in the online community. Media scholars have shown that the coverage of a traumatic event in the media evolves in stages over time, mirroring the social stages of meaning-making, from dealing with the loss to the assessment of cultural values and, finally, to the reaffirmation of group values (Dayan & Katz, 1992; Kitch, 2003). Many posts explicitly enacted a sense of national belonging, typically through presenting one's personal emotions of shock, grief and loss as feelings all fellow nationals share, but also through shared hatred toward the arsonists, as the following extract from a post shows:

> Stories will be told for the rest of our time
> Of a nation who reached out to appease the crime
> Crime which was punished but never justified
> For the lives that were lost, to our country, are tied.
>
> In our heart, in our eyes, we will feel the pain
> Not just of the lives, but the communities stained.
> Our Australian mantra of "She'll be right mate,"
> Spoken, but not felt, for our hearts felt hate.
>
> But Australians have an amazing ability
> To pull together in times of despair and hostility
> For the love of our people can not be compared
> We volunteer and donate because, put simply, we care.
> ("Our hearts felt hate," February 24, 2009)

As we can see, the contributions of ordinary people to the bushfire coverage on ABC Online represented a departure from the role they have traditionally played in disaster reporting, that is, as news sources or news characters represented by media professionals. Rather than being media-led, the affective community seemed to be in dialogue with itself, as the media provided a space for sharing emotions and forming a community.

Certainly, questions have been raised about the motives and moral implications of public expressions of emotion. For some, emotional self-expression encourages self-absorption that in turn distracts us from social engagement (e.g., Butler, 2004; Chouliaraki, 2006). For some, this confessional self-expression by witnessing audiences (and as we have seen, by journalists) can possibly be interpreted as a progressive moment in expressing and extending the boundaries of collective care and moral community. Robert Solomon (1997, pp. 244–245) made a convincing case for the virtue of public sentimentality in arguing that it "stimulates and exercises our sympathies without straining or exhausting them."

Conclusion

In this chapter we have conceptualized emotion as discourse, as a sociocultural phenomenon, through which relations of power are enacted and emotional meanings produced. We have discussed the appropriation and discursive deployment of emotions in disaster narratives and how they work to shape audiences' responses to local and global disasters. Disaster narratives give a public expression to the subjective emotions of victims, journalists, "ordinary people," and so forth—and it is through this representation that individual emotions enter the social and political realm. We have argued that rather than considering whether emotion is "bad" or "good" for disaster reporting or whether journalists should or should not express and evoke emotive reactions from their audiences, we ought to look at different ways and purposes of using emotion in reporting and consider its political consequences (cf. Seaton, 2005, p. 233). Therefore, the key question should be what emotions "do" in disaster reporting (cf. Ahmed, 2004; Harding & Pribram, 2009, p. 4; Reddy, 2001, p. 111). Academic studies and journalistic accounts depict emotion as a double-edged sword in disaster reporting: focusing (too much) on facts potentially diminishes the capacity of news to move people to action, whereas inserting (too much) emotion into a narrative often arouses critical concerns about sensationalism and exploitation of human suffering. Various scholars and practitioners argue for the need of "humanistic journalism" that is both affective and informational. While objectivity and empirical facts have been perceived as means to get at the truth, emotional content and self-reflection are increasingly justified in order to reach this goal. The understanding of the social roles of journalists covering disasters is in flux in the digital age, and in the journalism practice "truth-telling" and "witnessing history" seem to be more a matter of journalists making an effort to "shake" people than reporting in a disinterested, detached manner (Seib, 2002; Tumber, 2006). A cultural backdrop of an emergent affective public sphere and confessional culture in which personal revelations and introspection command wider interest in the mass media and popular culture may also be at work in encouraging processes of journalist reflexivity and inscriptions of care. We discuss this question in more detail in the next chapter.

PART II

MAKING DISASTERS MEAN AND POLITICALLY MATTER

5

Producing News, Witnessing Disasters

THIS CHAPTER SETS OUT TO EXAMINE THE PROFESSIONAL WORLD OF NEWS PRODUCTION and the current practices of journalists as they go about their work of producing news and witnessing disasters. It deliberately grants prominence to the firsthand accounts and testimonies of a range of broadcast producers, journalists and correspondents all working in and across the British television news industry, and all of whom have extensive experience producing news of recent major disasters. These insider perspectives afford an invaluable vantage point from which to better discern the complex of factors and forces at work and how these combine to select and shape news of disasters.[1] The seemingly systematic privileging by western news organizations and their journalists of some disasters as worthy of extensive and intensive media coverage, and yet conspicuous underreporting of others around the world, is well documented in the research literature (CARMA, 2006; Cottle, 2009a; Hawkins, 2008; Soderlund, Briggs, Hildeebrandt, & Sidahmed, 2008). This is explored further, as are the characteristic scripts and scenarios of disaster reporting and how journalists sometimes inscribe their reports with visceral scenes and emotional intensity aimed at "bringing home" something of the suffering of disaster victims and engaging audiences in their plight.

The idea of "witnessing" in the context of news about disasters and human suffering serves as our leitmotif throughout this chapter and provides the point of entry into the world of disaster reporting and journalist accounts of the same. Discussion of witnessing has recently received increased scholarly attention in the field of media and communication studies (Ellis, 2000; Frosh & Pinchevski, 2011; Peters, 2001, 2011; Zelizer, 2007). Scholars and researchers have usefully drawn attention to the concept's historical etymology and its origins in law, theology and atrocity that together endow witnessing with "its extraordinary moral and cultural force today" (Peters, 2011, pp. 24–25). Its inherent dualism rooted in its experiential ontology (the importance of "being there" and bodily "experience" of the events in question) on the one hand, and its epistemological dimension that depends on discursive representation (the act of testimony, narration and telling for others) on the other (Frosh & Pinchevski, 2011; Peters, 2001), has also been much commented on. This dualism, as well as the tensions that it can sometimes give rise to, as we shall see, is also immanent within the world of journalism and contemporary disaster reporting practices. An argument for regarding witnessing as central to the aesthetic sensibilities and cultural forms of modern means of communication generally has also recently been argued for (Ellis, 2000), though here we may want to draw a distinction between the seeming ubiquity of mediated 'eye witnessing' on the one hand, and mediated 'bearing witness' on the other with its added moral gravitas or injunction to care and, possibly, to act. Contemporary media and communications certainly position us daily as second-order witnesses to "media witnessing" of contemporary life in all its everyday ordinariness and mundane existence as well as moments of high drama, profundity and crisis. And, as we have already heard, recent scholarly attention focused on processes of audience reception (Höijer, 2004) and, importantly, how an ethics of care can sometimes be structured inside the form of news reporting (Chouliaraki, 2006) or become transformed in and through new forms of media and communication (Chouliaraki, 2010), point to further levels of complexity in media witnessing.

Barbie Zelizer observed how the centrality of "eyewitnessing" to journalism's professional self-image and claim to authority has shifted historically and specifically in relation to its crafting of news reports, journalism's professed role and use of available communication technologies (Zelizer, 2007). Across this evolution, she maintains, "eyewitnessing has highlighted ongoing issues within journalism about the most effective way to craft credible and authoritative narratives for happenings in the real world" (p. 425). The most recent evolution in new communication technologies and concomitant rise in "nonconventional journalists" chip away, she also suggests, at journalism's centrality as eyewitness to major events. These ideas are explored further in this chapter, with close reference to contemporary journalists and their accounts of what it is they do, and why, and how, when "witnessing" disasters.

Ideas and debates about witnessing *by, in* and *through* the media have yet to be pursued more explicitly into the production domain and in closer proximity to the professional practices and decision making of the journalists themselves. How today's journalists speak about and reflect on their own practices in the context of disaster reporting may yet help us to better understand how journalists, correspondents and news producers both conceive and perform witnessing with respect to major disasters and ground such practices within the organizational and cultural milieu of contemporary news production. To this end the chapter deliberately forefronts the words and accounts of the journalists and correspondents themselves and, on this basis, generates an opportunity to (re)assess the practices and contingencies now *at work* in the production of disaster news. We encounter from the newsroom point of view the geopolitics of disaster coverage (Chapter 3) and its infusion by "humanistic journalism" and "emotional discourses"(Chapter 4) and how each becomes enacted by the news journalists themselves. The rise of new communication technologies both within and beyond the world of news production and how these are now transforming the visibility of contemporary disasters around the globe (Chapter 2) are also followed up by similar means a little later on (Chapter 9).

As John Durham Peters (2011) reflected, media witnessing involves multiple dimensions:

> Media institutions have enthusiastically adopted its rhetoric, especially for nonfiction genres such as news, sports and documentary. Such titles as *Eyewitness News, See It Now, Live at Five,* or *As it Happens* advertise their program's privileged proximity to events. Media personae such as correspondents and newsreaders can be institutionalized as witnesses. Cameras and microphones are often presented as substitute eyes and ears for audiences who can witness for themselves. Ordinary people can be witnesses *in* media (vox pop interview, "tell us how it happened"), *of* media (members of studio audience), and *via* media (watching history unfold at home in their armchairs). The media claim to provide testimonies for our inspection, thus making us witnesses of the way of the world. (p. 23)

As we shall explore with the help of news producers, witnessing in the context of disaster reporting has indeed become deeply etched into the aims of news organizations and journalist practices and informs professional thinking and journalist reflections on what journalists do and why and how they do it. Ideas of witnessing have evolved historically, as we have heard, but in the field of modern journalism, there are grounds to suggest that they took root in the reporting of major conflicts and human disasters across the 20th century. Richard Sambrook (2010) observed the following:

> Foreign reporting in the 1930s and 1940s, like Gellhorn's and Murrow's, or the descriptions of the discovery of the German concentration camps from reporters like Richard Dimbleby, offered a narrative of conscience, a focus on victims and the humanitarian consequences of big events which still informs much international reporting today. They consolidated in the minds of editors and readers the core importance of bearing witness to unfurling events. (p. 6)

How news organizations and journalist practices "witness" disasters and enable audiences to do so secondhand (based on the news correspondent's witnessing) or thirdhand (based on news accessing of eyewitnesses) is not always a settled state of affairs—either across the news industry or for the individual correspondents and journalists concerned. "There is clearly an institutional politics of contemporary media witnessing," as Frosh and Pinchevski (2011) observed, "that informs how witnessed worlds are represented as shared and who may depict them and appear in them" (p. 11; see also Chouliaraki, 2010; Zelizer, 2007). Individual correspondents and journalists can often appear, as we shall explore, internally conflicted, entertaining competing views and values in their professional practices and declared aims and commitments. How news and journalist witnessing becomes organizationally and logistically accomplished, therefore, and how it becomes professionally enacted in the context of disasters can, in fact, reveal tensions and contradictions deeply embedded in the practices and ethos of contemporary disaster reporting. These tensions, we suggest, are literally and figuratively "at work" in the production of disaster news today, and they contribute to its characteristic forms, saliencies and silences as well as dynamics of change. One way of opening up these professional complexities and tensions to view is to first revisit the organizational logics and professional thinking that coalesce in the news industry's professional "calculus of death" (Cottle, 2009a).

The "calculus of death" encompasses the strategic choices and cultural outlooks of news producers that prioritize some deaths and some disasters as more newsworthy than others and that help to explain the conspicuous silences as well as notable inflections of much disaster reporting. Witnessing, in this context of journalists' priorities and engrained practices, often proves to be peculiarly stunted, selective and insensitive to the distant suffering of some people but not others. As we discussed in Chapter 3, based on current geopolitical interests and cultural outlooks, journalist interest in disasters in faraway places often becomes refracted through a cultural prism of ethnocentrism and Western interests. This has become institutionalized in the professional world of news production but as with the operation of general news values (Galtung & Ruge, 1965; Hall, 1973; Harcup & O'Neill, 2001), it remains taken for granted and generally unspoken by the journalists concerned. When invited to account for the selections and choices informing their disaster reporting, however, news personnel can readily elaborate on its informing tenets and shaping parameters. The discussion that follows, then, based on the first-hand accounts of disaster-experienced editors, reporters and correspondents working across the British TV news industry, first explores the professional calculus of death, how it is constituted in the world of news production and journalist practices, and how it impacts on the witnessing of disasters in different parts of the world.

Important as this engrained orientation to disaster coverage is, not least in terms of its seemingly cynical disregard for the moral equivalence of human life and human suffering around the world, there is, in fact, a more compassionate disposition sometimes at work in the field of contemporary journalism. This also demands recognition and exploration, though to date this has remained relatively unexplored in the research field. Also structuring the practices and performance of contemporary disaster reporting today is a thinly veiled but nonetheless transparent "injunction to care." This more performative journalist witnessing invites empathetic responses to the scenes and testimonies of human suffering, especially when brought home to viewers by correspondents and camera crews based inside the disaster zone and when witnessing close-up the scenes of human devastation. Such journalist witnessing can also sometimes lead to more critical forms of reporting based on the perceived failures of governments and/or those working in the field of disaster management and humanitarian relief. In these moments, disaster reports become communicated in and through the correspondents' crafted narratives, use of visceral images and embedding of voices that together inscribe their news reports with a morally infused injunction to care. How journalists reflect on, account for and manage this more performative witnessing opens up competing professional claims and commitments. The "injunction to care" does not always sit comfortably alongside the "calculus of death" or professional norms of journalism objectivity, impartiality and detachment. But, as we shall explore, there can be no doubting the journalistic inscription of an ethical commitment that now sometimes infuses news reports and invites audiences to care. This forms our second major theme of exploration pursued into the world of professional news production and practices.

The discussion that follows, then, unpacks some of the complexities of producing news and disaster witnessing. It does so by documenting the structuring antinomy currently at its core, that between journalism's entrenched "calculus of death" and its emergent "injunction to care."

From the Calculus of Death...

Journalists routinely and pragmatically make judgements about news stories and their relative importance and appeal for audiences. Particularly in the world of broadcasting, these judgements are informed by the limited opportunities of programme form, scope and available time for covering world events—including disasters (Berkowitz, 1997; Cottle, 1993, 2007; Wahl-Jorgensen, 2009). Calculations inevitably are entered into, a point generally acknowledged by journalists.

> I think in the BBC, as in other news organisations, there are decisions about how many people have died, how unusual is this kind of event. You know, there are calculations as to whether it gets on to the news agenda. (BBC Shanghai correspondent)

Logistics, Cultural Horizons and News Audiences

The calculations of whether a particular disaster secures a place on the news agenda are informed by a complex of factors and competing considerations. These include the perceived scale of disaster, questions of available resources and logistics, as well as the established positioning of particular news organizations in the wider news ecology (Cottle, 2006a; Cottle & Rai, 2006) and perceived audience interests. These, moreover, are not confined to the operations of news outlets but also inform their key gatekeepers: the news agencies. An Associated Press editor explained as follows:

> In terms of what gets covered and what doesn't get covered, you start to look at the kind of numbers involved and how bad, for example, this drought is compared with the other drought that we're covering. So, I would say that there is a bit of a numbers game. There is a bit of "how bad is it actually." I think that things have to be quite bad before the news media start covering them. There's not very much preventative cover, if you like. It's all at the point of crisis....And then there's a lot to think about—is it easy to get to and more often than not what's happening in the rest of the world, because you have a scant number of resources....So you make quite brutal decisions....And then, are your clients in their half hour bulletins going to run more than one African disaster that day? You know, a million people starving somewhere or dying of cholera, and then we have another little piece "meanwhile in Kenya they haven't had rain for a year," or whatever—it sounds so horrible—but we think they're not going to run that story so I'm not going to spend my money doing something that is not going to get any airtime. (Editor, Associated Press)

Logistical and resource considerations as well as the play of cultural outlooks evidently are at work in the selection and construction of news about disasters in different broadcasting organizations. Though a few journalists, such as the BBC correspondent quoted here, may sometimes suggest that disaster reporting and its relative salience and silences is all about logistical capabilities in space and time, most point to the play of other, more cultural, pushes and pulls at work.

> I think it is just simple logistics, you know, if there's an accident in China there's a good reason it's not covered: it's because no one's got access and nobody can really find out what's going on. And if 5 people get killed in a collapsed bridge in Wisconsin it gets covered because our 'copters and camera crews are all over the place. (BBC North American correspondent)

This was not, however, the common view encountered across the interviews. Though issues of resources and logistics are of course deeply implicated within the nature of disaster reporting and have yet to be thoroughly researched and theorized in comparative production studies of different disasters, according to the journalists themselves this does not explain everything. Journalists generally point to logistical *and* cultural forces at work when articulating their accounts of disaster reporting, and they are also capable of volunteering criticisms of how these sometimes combine to relegate some disaster stories to fleeting news mention:

> A lot of the time it has to do with the news cycle—what's around at the particular time. I mean, for instance, looking at what's happening at the moment with Libya, a lot of our resources are in Libya and moving around is obviously hugely difficult. Making sure you have all the equipment and the right personnel in the right places. If there were to be, say, an earthquake now in California, we would be severely stretched and that may hamper our ability to cover the story effectively. Sky News was criticised for not covering the Pakistan flooding (2010) as extensively as some other broadcasters and we are a smaller organisation—we did our best, we don't have anybody based in Pakistan. So you have logistical issues I think. With Haiti (earthquake 2010) they knew that they could get us out there relatively quickly, that there were flights that were still getting into the Dominican Republic and they knew they could get their Washington team there relatively quickly. So logistics do play a huge part, but it is incredibly controversial as to why. I can remember my father, an old journo, telling me when I was young about the one British person equals however many Bangladeshi etcetera. And obviously it's not as cold or as calculating as that, but you do tend, I suppose, to give more emphasis to things that are closer to home and that may affect more directly your viewership.…So logistics I think is one of the reasons but you know on other occasions I think it is worthwhile exploring why, because it's not always defensible. (Sky US correspondent)

This professional capacity to reflect on how cultural outlooks and cultural proximity shape and sometimes silence some disaster stories around the world also recognizes how others can find increased prominence by the same means. Speaking about the New Zealand earthquake that hit Christchurch on February 22, 2011, with the loss of 181 lives, two BBC journalists are in no doubt about the significance of cultural proximity or how it influenced reporting.

> Christchurch had a comparatively low death count for a disaster, but when you look at the amount of media coverage it got, and the number of human interest stories coming out of that, I think a lot of that is because we can relate to it. They're English speaking, they are the same race and although ethnicity in Britain is much more multi-cultural than somewhere like New Zealand, it's still predominantly a white, Anglo-Saxon, protestant kind of population than say somewhere like Pakistan or Afghanistan where we don't understand the people as well. (BBC news editor)

> Because the Christchurch earthquake was a first world disaster where, within a day, it was all closed off, it was very difficult to actually work there. And on the one hand, there were the residents saying "Gosh, the place is like a war zone" but most of the BBC people had been to war zones and it was nothing like that. It was just a few buildings that collapsed and a lot of cracked roads. But on the other hand, it was a place that looked like Guilford and so to a domestic news audience, you know, that is much more shocking than a landslide in the middle of China. (BBC Shanghai correspondent)

The cultural horizons informing the news producers are both shaped and informed by the perceived cultural affinities of their news audiences, and this becomes no less institutionalized in the operation of the calculus of death in particular news organizations.

> If the disaster is in a country like a third world country or developing world country or China or somewhere, that usually means that it will get coverage only if huge numbers of people have been killed or if it's something unusual like a volcano. (Only then) is the story perceived by the Editors to play big with the domestic audience. (BBC Shanghai correspondent)

Disaster Stages, Cultural Scripts, Iconic Images

News reporting of disasters has become typified in terms of its generally anticipated sequence of reporting stages, and, again, these have become consolidated into known cultural scripts helping journalists facilitate their reports as disasters unfold through time and as they practically negotiate the changing disaster scene.

> You know, every disaster goes through stages. Rescue first, then relief, then enquiry into what went wrong. And I think that most journalists covering a disaster go through similar stages. When you first arrive on the ground it's very much just what can you see, and then the next stage is, you know, can you go and find the people to whom this horrible thing happened and try to understand from them what happened and what was the impact on their lives. And then the third stage is people starting to ask questions as in the Christchurch earthquake for example: "What went wrong?" "Was there a problem with these buildings?" "Should they have been allowed, should they have been passed?"—that kind of thing. (BBC Shanghai correspondent)

Professional perceptions of audience links with the domestic context also influence the professional journalists' deployment of disaster scripts, how these become actively narrativized over time and inflected to appeal to national outlooks.

> Well, I think there is a formula, it doesn't always go in the same order necessarily, but there are key elements that you always have. So you always have your initial reaction, your initial report as to what's happened and then you have a mad scramble for information and then you have the main agencies, the main bodies coming out, you know, the Government or the relief agencies on the phone and then you start getting the facts and then you start getting the human interest. So like in 9/11, the people walked up to the cameras, and then you dig further and then you get the back story and then you get the reaction from you know, loved ones back home. So it is a certain flow, but then everyone is different because you don't know what stories you're gonna get and so the reaction is different. (BBC news editor)

Evidently more complexity is at work than simply the imposition of an invariant disaster script, pragmatically useful though this may be, and as journalists frenetically seek to make sense of and report on the onward rush of disaster events.

The nature of different disasters also plays a part in determining how they will become reported over time. Here a BBC correspondent reflects on the BP Deepwater Horizon oil spill in the Gulf of Mexico that killed 11 people and spewed 53,000 barrels of crude oil into the marine environment every day from April 20, 2010, until July 15, 2010—the largest oil spill in the history of the petroleum industry:

> The challenge with the oil spill was we knew it was huge: you saw these aerial pictures, you saw the rig on fire. But the problem with that one was that oil didn't come washing up on shore, people weren't seeing drowned birds and dolphins smothered in oil—it wasn't like *Exxon Valdez*. It was so hard to visualise that. So that had its own challenges. But then you started to see after a few weeks how it was affecting restaurants in New Orleans or oyster fishermen who were fourth and fifth generation oyster fishermen—and that actually stayed in the news for a very long time, for different reasons: one, because it was a

> British company—lots of shareholders, lots of pensioners had their money invested in it, and two, it also affected millions of peoples' lives along four states and the Gulf coast.... But then you can only do oyster fisherman and restaurants and peoples' lives so many times before it starts to get repetitive and you have to find fresh angles all the time to try and keep this story in the news. But ultimately, these stories will eventually fade away because they'll be replaced by something else. But you know, the one-year anniversary of the oil spill is coming up and I know I'll do a lot more stuff on that again and probably in years to come as more and more results come through as to the environmental impact. But logistically that was a tough one to do because you couldn't see much. You had these images in your mind as to what you think it should look like, but it didn't look like that.... (BBC North America correspondent)

In the case of the BP oil spill, then, we hear of the failure of this spillage to conform to established expectations based on prior oil spills at sea and notwithstanding the fact that it was the largest in world history. Here the preceding template appears to have been the *Exxon Valdez* oil tanker spill in Prince William Sound, Alaska, on March 24, 1989, that discharged up to 75,000 barrels of crude oil. Unlike the *Exxon Valdez*, or the *Torrey Canyon* tanker spill off the coast of Cornwall in Britain before that in 1967, the Deepwater Horizon disaster didn't produce the same culturally resonant and iconic images and evidentially exhibited other differences that also departed from the established cultural script of oil spillages at sea.

Some disasters, by their perceived nature, also produce more dramatic scenes and exciting narrative scenarios than others, once again pointing to the cultural attribution of meaning and significance to disasters. In the case of television news especially, some disasters, clearly, resonate more than others with the medium's predilection for live, dramatic images and breaking news.

> I think the thing about earthquakes is that it's not over straight away, because there's the chance of drama because people could still be alive. So it makes sense to go there as oppose to say an explosion in a mine in China or somewhere where often everyone is dead—you know, what do you actually gain from going? Whereas, with something like Haiti or Christchurch, there's the chance of more unfolding drama. The classic example, I mean it's not a disaster on the same scale, but the example of the Chilean miners—that's a perfect disaster in a way. You know, where there was the chance of rescue with some risk of failure. I think that's what had everyone hooked. (Channel 4 News reporter)

The search for iconic images and dramatic scenarios, including those already established by the media in earlier disasters, can assume a major imperative in the visually centric medium of television news. Sometimes, as in the Haiti earthquake of

January 12, 2010, where, according to the Haitian government, up to 316,000 people lost their lives, an inordinate amount of news exposure was devoted to the "miraculous" rescue of a handful of survivors pulled from the rubble over the following days. And some disasters, such as the 2010 Copiapó mining accident mentioned earlier by a reporter, where 33 Chilean miners were trapped 2,300 feet below ground for 69 days before rescue, can, by the same means, become catapulted by the international news media onto the world news stage. As the miners were finally winched to the surface on October 13, 2010, an estimated one billion or more people around the world watched the event live on their television screens. Disasters that can produce dramatic scenes of rescue and potent images of human endurance and heroism, then, can become propelled by the international news media into global media events. Those that do not exhibit such affinities with established cultural scripts, on the other hand, may go relatively unnoticed by the world's news media.

Political Context and News Ecology

Political context and forms of political control in disaster situations also condition news organization responses to different disasters and, in the case of television news, the production of "good television news pictures."

> I think that in China the biggest issue is whether we can get to it in time to be able to produce the kind of coverage that we want to. Now what I mean by that, if there is, say, an earthquake and it's in a remote place we have to be aware that it would take probably a day to get there. Now in China, the later you get there, the more chance that the authorities close it down and you can't get to it. That, with an earthquake, it usually takes them several days, because they've got to kind of gear up to it. The worse the earthquake or landslide, you know, the more difficult it is for them to get to that stage. So, with an earthquake or landslide if there is significant loss of life, then we will probably go because we know that if we invest the time and money in going there, then we will get something out of it. When you have something like floods for example, floods are difficult because even if there is significant loss of life and so that is judged to be a news story, floods come up and then they go. So there is a chance that it's harder to get good television pictures of what's going on. So last year (2010), there were floods all over China and there was a sort of drip, drip of bad news stories about people dying in floods but it wasn't really covered because it was so difficult to be in the right place at the right time when it happened. Whereas, something like the big landslide in Gansu (7 August 2010), that I went to, it was horrible. It was a huge stream of mud that came down the hill, engulfed the hillside, killed hundreds of people and it was obvious where it was, how to get there and so that was the one we went to and covered. (BBC Shanghai correspondent)

The same correspondent also draws attention to how disasters are stage-managed by political authorities, including the deliberate production of images and scenes thought to have propaganda value. The example of the Gansu mudslide of 2010 is followed up here, where elite media appearances were stage-managed and designed to demonstrate political control and personal sympathy in the face of the destruction and loss of life. Not that this is confined to authoritarian, nondemocratic regimes. Similar public relations efforts are also much in evidence in Western disasters where political leaders also feel compelled to personally visit disaster sites, hug survivors, congratulate emergency service personnel and make statements of sympathy and forthcoming action—in front of news cameras.

> There's very much a sense that governments now know that how they respond to a disaster is incredibly important. When I went to the Ganzu landslide, it was like a movie set—there were mechanical diggers carrying soldiers in the scoop, getting them into the place that they needed to be. And one of them would have a huge flag pole with a huge red flag like something out of the Cultural Revolution and this was because disasters were an opportunity for the Communist Party to promote the narrative that the party looks after you, the party rescues you. The Ganzu landslide had been caused by logging which had stripped the hillside of the trees. They didn't want people to focus on that. The Chinese Prime Minister immediately arriving by helicopter as only you can in a disaster and then being filmed for the national news on his hands and knees in a shed calling out to the people that are being rescued and saying "Don't worry, don't worry" and taking a very personal charge of it—the action man in charge. And that's how they see disasters now. The Communist Party see that there's an opportunity, a propaganda opportunity. But when it becomes nasty, like in the Sichuan earthquake, as soon as people started asking questions about why all the schools had fallen down, well then they tried to shoo the media away. You know, they've learnt probably from the Western PR advisers that the best thing to do, the best take on very bad news, is to get journalists completely obsessed with one job that they are doing so they don't have the time to ask the questions about why the schools had fallen down. (BBC correspondent)

When situated in political context, then, disasters become opportunities for elites and political authorities to reassert their claims for legitimacy and control, but they can also become moments of political crisis, scandal or even the mobilization of political opposition (see Chapters 2, 7 and 8).

A further consideration informing the operation of today's professional "calculus of death" is the differentiation found within and across the contemporary news ecology. Whereas BBC news operations and journalists, for example, are traditionally disposed to stress the corporation's professed goals of impartiality and objectivity in disasters and the dissemination of accurate information about them around the world, both Sky News and ITV news, in contrast, are more inclined to emphasize their strengths in rendering disasters culturally meaningful for their particular audiences; and Channel 4 News, according to one of the reporters interviewed, takes pride in finding innovative news angles and news stories not normally pursued by the other major news outlets.

> It depends on the outlet you are filing for and the story that you are covering, but there's no doubt about it that each individual news organisation will bring its own flavour to it, and I suppose Sky will go for a much more human interest angle if it can. We film and we report on stories that we think would be interesting to the Sky audience and tell it in a way that we think would appeal to them. So with the BBC it's very similar—for the World Service you would go into much more specific detail perhaps and assume much greater knowledge than for a Sky viewer. So you have to take that into account. (Sky US correspondent)

The differentiation within the field of broadcast television news (Cottle & Rai, 2006; Matthews & Cottle, 2011), and indeed within the wider ecology of news, also impacts on the operation of the calculus of death and how it is applied by particular news outlets to disasters from local to global levels. This deserves closer inspection. What we have begun to document here, however, is something of the shared and underlying forces now at work—logistical, practical, cultural and political—that inform editorial and journalist selection processes and shape general disaster reporting. These, as we have heard, coalesce in delimiting and conditioning not only the selection of disasters reported from around the world but also how they become culturally signified and made to mean. This can produce a peculiarly myopic and amoral—if not immoral—"witnessing" of disasters, death and dying, as we have heard. But this is not the whole story. Attending to the professional world of news production and journalist practices reveals a different professional disposition also widespread in the British news media industry, one that exists alongside but also in some tension with the pragmatically expedient calculus of death described. This is documented and discussed next.

...to the Injunction to Care

Journalists today are capable of doing far more than simply recording the latest death tolls and reporting the basic facts of selected disasters based on available logistics, cultural outlooks and the morally evacuated calculus of death elaborated earlier. Journalists and correspondents also produce news packages and craft their stories in sometimes powerfully culturally resonant and affective ways, as we discussed in Chapter 4. They produce their news stories to appeal to audiences and to engage them—cognitively, affectively, culturally and, sometimes, morally. This, inevitably, raises questions about the nature of contemporary journalism witnessing and the possible tensions inherent within this in respect of traditional commitments to journalist impartiality, detachment and objectivity. Evidently some journalists feel, as we shall hear, that their personal and professional commitments exist in some tension with established journalistic expectations of professionalism.

Engagement Beyond Death Tolls

Though entertaining the idea that their reporting of disasters and human suffering may elicit responses of care from their audiences, professional claims to impartiality and detachment generally keep such views in check, inhibiting the explicit advocacy of how audiences should feel or respond. Adherence to these core tenets of professional journalism remains strongly in evidence across the mainstream field of broadcast news notwithstanding calls for alternative or "corrective journalisms" from "development journalism" and "peace journalism" to "public journalism" and the "journalism of attachment" across recent years (Cottle, 2006a). This is not to say, however, that mainstream broadcast journalists do not want to engage their audiences or do not deliberately set out to do so.

> I want them to engage. I'm not necessarily going to tell them how they should feel about it, but I want them to be engaged in the story. Whether they think "Why are we bothering about that" or "I need to reach into my wallet and give them £5"—you want to get some kind of reaction from your audience. You want to stimulate their interest. So, do I want them to care? I suppose if I step out of my reporter's suit as a human being, sending back pictures we have filmed that were profoundly upsetting, I would want them to be concerned, yes. But as a journalist, I can't tell them to be concerned. (Sky US correspondent)

Encapsulated in this statement of professional intent, then, is the conundrum of wanting to make sure audiences "engage" while eschewing any suggestion that the journalists have set out intentionally and deliberately to make audiences care, for that would imply they have crossed the line of journalist professionalism and its demarcations of impartiality, detachment and objectivity. Even at the BBC, with its historically forged and institutionalized commitment to these core tenets of journalistic professionalism, normative ideas of caring remain barely concealed in journalists' thinking, especially when trying to make sense of the horrific images and accounts of disasters that can interrupt and unsettle everyday life.

> It's a very fine line. It's a very difficult one to walk, I think. Because on the one hand, you do want to be impartial and I don't want to tell you to feel desperately sorry for those people in Christchurch or for those people that were on the beach in Thailand when the tsunami hit on their honeymoon, or whatever the story is for that family in Eritrea that has had everything decimated. But at the same time, you want people to really care, because, you know, life is not a rehearsal and you should care. While you're complaining because your McDonald's is a bit too hot, there are people dying that could be saved by the £2 you spent on that coffee. (BBC editor/reporter)

Some news reports, evidently, set out to do more than simply document the events, consequences and responses to disasters, important though this may be to the historically forged mission of news journalism. News organizations and journalists also want their reports to resonate with audiences and emotionally "hit home." Here a correspondent reflects on how his popular variant of TV news permits and indeed requires that stories of disasters move beyond the bare facts of disaster and deploy ways of humanizing the suffering involved to bring this home to audiences.

> Yes, I mean we have a very clear idea of our audience and I think that colours a lot of Sky News' reporting. We are unashamedly populist: we have a brighter, brasher approach than perhaps the BBC, but there is also that mission to explain, which is a core ethical value, I suppose. But what we have to do is, these people may look different from you, they may be living in different circumstances/situations than you at home, but in the end they are people too and you have to introduce therefore an individual who may be suffering and try to explain their circumstance, to make it resonate with the viewers at home. Otherwise it does become 20,000 dead—you start to fall into almost disaster porn, where the figures become the story and I think that's what responsible news organisations have to try and do, is to see beyond those numbers and try to make the story hit home. (Sky US correspondent)

This declared professional intention to make the story "resonate with the viewers" and "try to make the story hit home" is not confined to particular variants of popular TV news. Though populist forms of news are generally disposed to seek out "human interest" material and will often do so through deliberately seeking personalized and affecting stories (Dahlgren & Sparks, 1992; Cottle, 1993; Sparks & Tulloch, 2000; Lule, 2001; Langer, 2003), in the context of disaster news, as we shall see, this now represents a more general if not universal approach to disaster reporting. Journalists are apt to stress, nonetheless, the experiential ontology of witnessing and not their part in the epistemology of witnessing that requires active narration and re-presentation of testimony, even when the latter is clearly at work.

> I actually just try and simply paint a picture and be people's eyes and ears. And I really think it is just that simple. I don't think there's any political agenda; I simply try and reflect and take people from their homes wherever they are in the world to where you are. I think that's probably what I'd like to get across more than anything else. I just think we're on the ground and not doing anything else except trying to tell people what's really going on and trying to bypass all the bullshit really, bypass all what the Government people are saying or what the big companies are saying and get past all that and get down to the kind of David versus Goliath type scenario where you're just talking about who has really been affected and what's happening to those people and what the future looks like. (BBC North America correspondent)

Here a correspondent at the BBC, an institutional bedrock of journalistic norms of impartiality, detachment and objectivity, declares his principal aim to be the people's "eyes and ears" and simply document the impact of disasters on the people directly affected and yet, simultaneously, intimates how a preconceived cultural script may also be at work when seeking out the "David versus Goliath type scenario," a scenario that privileges the impacts of disasters on the lives of ordinary people. The slippage between the *experiential ontology* of witnessing that underwrites claims to be the "eyes and ears" of disaster and the *epistemology of witnessing* based on active narration and the structuring nature of established cultural scripts is barely concealed in such views (see also, Zelizer, 2007). News organization and journalist witnessing of disasters involves bearing witness to human trauma and tragedy and thereby helping audiences to recognize the human plight of those caught up in such catastrophic events. This witnessing and inscription to care are conveyed within the textual forms and crafted narratives of news storytelling, even though the professed norms of impartial and detached reporting ordinarily demand that explicit entreaties to care are not permitted under the taught conventions of professional journalism.

The opportunity to humanize news stories and populate them with individual accounts and personal testimonies aimed at "bringing home" the human reality and suffering of disasters depends in large measure, according to the journalists, on the news organizations' logistical capability to position reporters inside the disaster zone, "to be there" and not simply rely on agency materials.

> These days we can get coverage from Reuters and APTN and if we don't have a man or woman on the ground, then you're really just reliant on those news agency pictures. But if you're writing to somebody else's pictures you tend to keep it more factual; whereas if you want to go to one of those disasters, if you've made the effort to go there, news is about the people, it's not about events. So there you want to try to explain to people what the impact of this particular disaster has been on individuals, and so your first journalistic instinct I think, as a broadcaster, whether it's TV or radio, is to go and find somebody who

> has a good story to tell. It's impossible to tell the impact of an earthquake on a village if the whole village has been flattened and one of the most effective story-telling ways to do it is to find an individual and say "This is Bob, and Bob was sitting in his house and the earth fell in on him and his whole family disappeared." (BBC Shanghai correspondent)

Though some may want to question why only journalists and correspondents "parachuted" in from the major news organizations can be effective witnesses given the rise of new social media and informed local journalists often available on the ground (Pedelty, 1995; Zelizer, 2007), there is no doubting the continuing journalistic importance attached to reporting from the scene and narrating and imaging what has happened close-up.

Bearing Witness from the Disaster Zone

Reporters and correspondents on the ground witness disasters and their aftermath firsthand; they see terrible scenes, hear traumatic accounts, smell putrefying corpses and bodily encounter and experience the conditions of destruction that engulf countless victims and survivors in a desperate struggle for survival. This professional "embedding" within the disaster zone lends credibility to the correspondent as witness and by association to his or her news organization (even though historically and theoretically we know that "being there" does not guarantee privileged insight or infallible knowledge). This "being there" nonetheless underpins the primary ontological claim of bearing witness. A foreign correspondent makes the case when contrasting "being there" with the technologically facilitated "connectivity" that new media and communications now permit.

> To bear witness means being there—and that's not free. No search engine gives you the smell of a crime, the tremor in the air, the eyes that smoulder, or the cadence of a scream....I have been thinking about the responsibility of bearing witness. It can be singular, still. Interconnection is not presence. (Roger Cohen, foreign correspondent, cited in Sambrook, 2010, pp. 91–92)

Their "standing within the field," both literally and figuratively, is professionally thought to enhance the authority of foreign correspondents and reporters to document and describe what is going on. As in other zones of civil destruction and trauma, whether conflicts or wars, for example, this proximity to events can also lead to a deeper understanding and felt commitment to communicate to others the plight of those whose suffering has been personally observed. And so too can this immersion in the field sometimes give rise to a sense of frustration or anger when humanitarian needs are perceived to remain unmet. The emotion of anger often exhibited by disaster survivors, their families and wider communities and directed at those thought to be in some way culpable or, subsequently, incompetent (see Chapter 8), can also potentially inform journalist responses—notwithstanding the felt professional necessity to keep these and other emotions carefully in check, as we discussed earlier.

> Well, personally for me, it's because I care and, unless you've got a lot of empathy, you can't really be a reporter. I think for me and for a lot of other people that covered Haiti, you know, it was traumatic for us. We saw things that you wouldn't want to see in your worst nightmares. And it makes you care you know. We're supposed to be objective, but of course you care about people and you want to keep that in people's minds—that this is what happened, it was important, a lot of people died, but it's also important that you see that the money has been spent properly, or it's not been spent properly or just try and keep it in people's minds so that they are still aware of what happened, because people forget so quickly. (BBC North America correspondent)

From this embedded vantage point on the ground, witnessing firsthand and hearing the accounts of others, the disaster correspondent secures enhanced professional autonomy to take a more critical and interrogative stance to the management of disaster relief. A correspondent based in a competitor news organization also draws attention to this opportunity in the Haitian earthquake of 2010.

> Why weren't the supplies getting to the people because we were seeing huge stacks of supplies coming from all over the world being piled up at the airport and I think the British media on the whole did quite a good job of saying "why isn't this stuff getting out to those in desperate need just the other side of the barrier?" And a lot of it was to do with organizational concerns and fears that unguarded UN conveys could come under attack. I mean, I think they were unfounded fears because these people were just desperate. But that was a story that we felt we were seeing in front of us unfold and we had to tell it. And so, it changed. (Sky US correspondent)

Such witnessing of and bodily exposure to the precarious conditions of existence left behind in the aftermath of the Haiti earthquake help in part to account for the critical attention directed by the correspondents on the ground toward the humanitarian relief effort and the perceived failure of the UN in particular to effectively coordinate and distribute humanitarian supplies. Processes of journalist immersion within situations of disaster and humanitarian emergency (Cottle & Nolan, 2007), as well as situations of political conflict such as the Arab uprising of 2011 (Cottle, 2011b), point to a more beneficent form of journalist "embedding" than that discerned in the context of war reporting (Morrison & Tumber, 1988; Morrison, 1994; Pedelty, 1995; Tumber, 2004), but it may be no less informed by the dramatic unsettling of civilian norms and bodily exposure to conditions of devastation and human loss. In the potentially traumatizing conditions witnessed by the correspondents in Haiti, their capacity to form supportive networks and friendships further helped to sustain their sense of group identity and affirm their empathetic and critical responses to the unfolding disaster and its unmet humanitarian needs.

> Logistics were difficult out there and Channel 4 was struggling for a place to sleep. And we're friends—although they were the Washington Bureau—we all know each other quite well—from a spiritual point of view as well: because you're working so closely with your team and spending a lot of time with them frankly, in pretty nasty conditions. We were lucky to be in a hotel with a working bar with a very large cellar—sitting and talking through what we have just witnessed with people who weren't just in your team and were there and sharing those experiences. I know some people did have some issues coming back from Haiti afterwards, and were very upset by what they saw. We seemed to fare better than most, and I really think part of it was that there were all these different teams all sitting around and all talking to each other...(Sky US correspondent)

Journalists involved in disaster reporting, as we have heard, profess to care, and they now have increased opportunities to demonstrate this more compassionate side publicly, in the surrounding world of blogs and the circulation of introspective pieces specially written for other news outlets based on personal reflections and experiences of disaster reporting. Their professed hope that audiences and readers may also care, based on their news packages and images communicated from the disaster zone, is one further way in which this compassionate stance can find a legitimate outlet while also maintaining adherence to claims of impartiality, objectivity and detachment.

Bodily Senses, Privilege and Moral Obligation

The previous discussion has not sought to suggest that all journalists simply seek to *sensationalize* their news stories in the sense of filling them with emotive images and the traumatized accounts of survivors. Journalists, for the most part, vigorously rebut charges of "sensationalism" and reflexively respond to them in their reporting practices.

> I don't really ever put a piece together thinking this is going to make people cry, but there are definite times where I've actually wept myself as I'm putting them together, because it's simply just so moving. And if I have that reaction, I think you know, I hope that other people might have a similar reaction. But all I'm trying to do is actually take them there, you know, mentally take them to the school I turned up to on day two of the earthquake in Haiti where a woman walked up and found her son dead on the ground or where there

> were little children's hands and feet sticking out of the rubble. I mean, that wasn't me trying to make people cry, that was me just literally getting out of a van, walking up with a radio mike and describing what I could see as best I can to paint that mental picture and take people there. I'm never really thinking this is gonna make Mr and Mrs Jones cry their eyes out and pull a cheque book out. I felt in Haiti in particular it was a privilege for me to be there, to actually have the chance to tell the story. (BBC North America correspondent)

It is evident that journalists' witnessing emphasizes the observational, the privileged "I was there" mentality of disaster reporting, a position that registers the primary sense of witnessing in terms of its bodily and sensory experience of being at the scene of destruction (Peters, 2011; Zelizer, 2007). This presence can also produce a commonly articulated sense of witnessing as "privilege," as we have just heard, a sense in part that refracts its earlier religious antecedents and moral connotations. At least three mutually reinforcing senses of the "privilege of being there" are at work here. First, the privilege of "being there," witnessing and yet escaping whether by luck or circumstance the terrible fate that has just befallen others; second, the literal privilege, perhaps, of occupying a high-status, well-paid position in the comfortable "first world" that becomes all the more "felt" when reporting on those who, often in the "third world" and experiencing disaster, have nothing or little left in their daily struggle for existence; and third, and importantly, the privilege of bearing witness that gives rise to a sense of obligation to those whose plight and suffering have been observed and that now silently command that this must be communicated to others, because the dead and vulnerable are not in a position to tell the world for themselves. This sense of moral responsibility to record and impart to others, to "bear witness" to pain and loss and communicate this to the wider world, echoes not only historically forged ideas of "witnessing" (Peters, 2001), but also journalism's own contribution to this with the reporting of war, atrocity and humanitarian disasters across the 20th century (Leith, 2004; Sambrook, 2010).

If the experiential ontology of witnessing lends authenticity and possibly moral gravitas to the reporters' accounts and testimonies, it is the epistemological crafting of these acts of witnessing by them into effective narratives replete with dramatic images and heartfelt voices and accounts that engage audiences and help them recognize the human suffering involved. The correspondent's declared sense of "privilege," a phrase often used by reporters when reflecting on their role in witnessing destructive human events, and one that transcends simply "eyewitnessing" and "being there," registers this added sense of moral gravitas and obligation. The latter becomes discharged, in part, through communicating to the wider world what has occurred, a moral as well as a professional responsibility to those whose lives have been violently decimated and disrupted, and a communication act that is invariably informed by the hope that it can help to sustain bonds of moral connection and responsibility on the part of "witnessing" audiences.

There is more to the journalistic injunction to care in contemporary disaster reports, therefore, than simply the manufacture of "sensationalism" or the deliberate inclusion and profusion of emotive accounts and traumatic scenes. Correspondents on pragmatic as much as professional grounds generally eschew sensationalism.

> So I don't think it's particularly believable frankly when a reporter racks on about how everyone's feeling. My thing would always be that it is for the people you interview to say, it's not for you to say and also I just think as a reporter there's more important things you could be saying, so leave the emotion to the people who you are interviewing. But also, I think, the most powerful piece of emotion is often when your script is quite spare or not overwrought because you just let people say what they're going through. (BBC reporter/editor)

In other words, journalists' injunction to audiences "to care" is typically encoded and enacted in and through their news packages based on carefully crafted narratives, use of visuals and the positioning of different views and voices. It is through these and other forms of textual inscription that journalists can hold onto their professional claims to impartiality and detached journalism while nonetheless also discharging their felt obligation as "witnesses" to other people's suffering.

If journalists generally seek to ward off their critics and protect their claims to impartiality and objectivity through the "strategic ritual" of accessing authoritative sources (Tuchman, 1972), so too are they no less professionally accomplished in crafting and visualizing their news stories with a less than explicit but nonetheless transparent injunction to care. There can be no doubting the inscriptions to care infused in many disaster reports or how these critically depend on the craft of journalist writing and visualization. This, as we have already heard, can be crafted in part through the deliberate use of visceral images (whether based on words or film) and embodied in the evocation of bodily senses of sight, sound, smell, taste and touch (invoking *senses* rather than gratuitous or base *sensationalism*). As one BBC correspondent said about the Haiti earthquake, "In Haiti, I felt like it was a privilege and you just want to paint a picture for people so that they can smell and hear and see and feel what you are on the ground." This experiential ontology of "witnessing," of bodily being there, is here discharged through the deliberate invocation of bodily senses. Much depends of course on the medium concerned, whether live or recorded TV, radio, press, on-line and so on, but the inscription of bodily senses can generally be applied within different media when seeking to engage audiences. It can also become a potent device for discharging the felt moral obligation of bearing witness to those whose suffering and tragedy has been observed. An illustration helps to demonstrate the point. In the following BBC news report, for example, different human senses deliberately feature in and across the reporter's narration. And here the act of journalist witnessing, of "being there," becomes epistemologically dependent on its representation through the news medium concerned. It is in the moment of disaster narration and storytelling that the powers of the journalist's craft become instantiated and inscribed inside the act of news witnessing—sometimes tellingly so.

> **Haiti desperate for help after quake**
> *By Andy Gallacher, BBC News, Haiti*
>
> The situation here in Port-au-Prince is now at a critical point, with rotting corpses beginning to fill the streets.
>
> The cries of help that were being heard from the rubble have now been silenced—for many people it is simply too late.
>
> Haitians feel very alone at the moment. The promise of aid has not yet materialised and many locals are still digging through the rubble with their hands.
>
> Most of the bodies are covered in white bed sheets or rolled inside carpets, but others have been left exposed to the hot sun and the stench of rotting bodies has begun to fill the air.
>
> Families who are desperately searching for their loved ones are gingerly uncovering the sheets that cover the corpses in the hope they can at least identify family members.
>
> But even if bodies are identified there is nowhere for them to be laid to rest.
>
> Mass graves are now appearing across the city.
>
> The mood for the past 24 hours has been one of patience and solidarity, but there is now a sense of anger and frustration that could change the atmosphere here drastically.
>
> "This is not the time to blame anybody. This is a natural disaster, only God knows why it happened," says Louinel Staibord, who came to Port-Au-Prince from Florida to find his family.
>
> "I believe that this is the time where everyone should help each other, this is a time for generosity, we should sympathise with each other."
>
> Schools hit.

Louinel is one of the lucky ones, he has now found all his family members alive and well.

"Haitians can only depend on international help because the infrastructure here is decimated."

***

The rescue effort and the promises of help are now desperately needed, but so far the fresh supplies of water, food and medical equipment are still in short supply.

Some of the worst hit buildings were schools.

Several had more than 1,000 pupils inside when the massive earthquake hit, and there is little left but concrete blocks piled one on top of another.

The bodies of children and adults can clearly be seen, and most will remain that way until the rescue teams get on the ground.

At night Port-au-Prince grows eerily quiet.

Most people are still too afraid to take shelter inside the buildings. Tremors are still being felt here and even the hospitals are treating their patients in the grounds.

The airport, rapidly becoming the centre of this rescue effort, is in full working order.

The building was damaged but there is power and the runway, despite some reports, is in good shape.

Military planes are landing more regularly than they were last night, but there is no sense that the operation has begun in earnest.

People can now only hope and pray that help will arrive. Haitians can only depend on international help because the infrastructure here is decimated.

Time is not on the side of the Haitian people; help is needed and for many it has already arrived too late. (BBC News online, Page last updated at 01:49 GMT, Friday, 15 January 2010) (http://news.bbc.co.uk/1/hi/world/americas/8460547.stm) (accessed 7.12.11)

This news report inscribes an injunction to care, but it is not carried for the most part, as we can see, through explicit advocacy, emotionally charged statements or direct exhortations for humanitarian intervention; neither is it discharged through the deliberate positioning of a series of emotional testimonies from others. Rather, its inscription to care is enacted through the news narration itself, in its succinct but sensory-invoking images and emotionally detached analysis of what has occurred and what needs to happen next. It is by these means that the piece underwrites its central proposition: "Haitians can only depend on international help because the infrastructure here is decimated" and "Time is not on the side of the Haitian people; help is needed and for many it has already arrived too late." The sense of sight, for example, is invoked through a series of graphic images: "Most of the bodies are covered in white bed sheets or rolled inside carpets," "The bodies of children and adults can clearly be seen," and "Many locals are still digging through the rubble with their hands." Invocations of sound and hearing are also crafted into the news report: "The cries of help that were being heard from the rubble have now been silenced," "At night Port-au-Prince grows eerily quiet." Visceral references to smell also feature: "The stench of rotting bodies has begun to fill the air," as do proxy indications of touch: "Digging through the rubble with their hands," "Others have been left exposed to the hot sun."

In such crafted, sensory ways, then, journalists live up to their ambition to "paint a picture for people so that they can smell and hear and see and feel what you are on the ground," grounded in the ontology of witnessing as "being there." As they do so, however, they simultaneously demonstrate the epistemological potency of the act of narration and telling, of giving testimony, discharged through their crafted prose. The only voice directly embodied in this piece, other than the journalist's, is also not confined to a personal situation or feelings but simply advances a moral call for what should happen next: "I believe that this is the time where everyone should help each other; this is a time for generosity; we should sympathise with each other." The journalist finally concludes the report by lending endorsement to this moral claim by repeating the earlier emboldened statement: "Haitians can only depend on international help because the infrastructure here is decimated," and by reminding the reader that "Time is not on the side of the Haitian people; help is needed and for many it has already arrived too late." This, then, is much more than a straightforward description of a disaster; it is a carefully crafted elicitation, an inscribed "injunction to care."

In such reports of news disasters, we can often find similar journalistic inscriptions. How these variously become enacted and performed by journalists today demands in-depth exploration and analysis not only in respect of the textual representations of disaster (Chouliaraki, 2006, 2010) but also in respect of the world of news production and its professional practices. The BBC article discussed above simply helps to indicate how journalism, notwithstanding the news industry's continuing adherence to institutionalized logics and cultural outlooks that coalesce into a professional calculus of death, now also sometimes seeks to discharge a different, and more compassionate, set of professional aims and commitments.

Conclusion

It is clear from the testimonies and accounts of the journalists granted prominence across this chapter that a professional antinomy is currently *at work* in the reporting of disasters—the institutionally naturalized and professionally enacted calculus of death on the one hand, which facilitates logistical deployments and the prioritization of some disasters, some deaths, as seemingly more morally significant than others; and a thinly veiled injunction to care on the other, that is professionally crafted and inscribed into the narratives and visual scenes of disaster reporting. News witnessing of disasters, evidently, is subject to processes of institutional mediation, cultural inflection and professional reflexivity. Bearing witness in the field of disaster reporting carries the traces of all these, and correspondents and journalists negotiate a sometimes difficult path between them as they seek to live up to different expectations and felt obligations. How this deep-seated antinomy in the world of news production and disaster reporting changes in the future and how, if at all, the weighting between the professional calculus of death and the injunction to care may shift will depend in part on wider processes of cultural change and today's rapidly evolving communications environment.

Journalists today, as we have heard, professionally recoil for the most part from any suggestion that they can or should seek to convey personal views, display their emotions or instruct audiences on what to think or how they should feel—even when confronting the humanly destructive consequences of disasters. And yet, the tradition of bearing witness crystallized in the tradition of foreign correspondence imbues, it seems, the role of disaster reporter with a sense of "privilege" and associated sense of moral obligation. And both, as we have also heard, can be keenly felt and professionally inscribed within journalists' news reports. How they negotiate these competing professional expectations and felt obligations in practice reveals a more complex state of affairs than is sometimes assumed in the research literature, and one in which some disasters, on some occasions, are reported with a professionally crafted injunction to care. This, as we have seen, is discharged in and through the professional craft of writing, narrativizing, visualizing, populating and personalizing news stories, and it secures its warrant from the correspondent's institutional and ontological positioning as witness, being there, in the field of disaster. How this institutional and ontological positioning is evolving in today's rapidly changing communication environment is a discussion we return to in Chapter 9, when we consider the role of new communication technologies and social media in the transformation of disaster visibility. How journalism's injunction to care becomes complexly infused with discourses of cosmopolitanism and nation now forms the subject of the next chapter.

Note

1. This chapter (and parts of Chapter 9) are based on interviews and follow-up discussions with the following media personnel: Brendan Bannon, photojournalist; Jane Deith, Channel 4 reporter; Matthew Eltringham, BBC associate editor, Interactivity and Social Media Development; Deb Evans, BBC news editor; Andy Gallacher, BBC—North America correspondent; Chris Hogg, BBC—Shanghai correspondent; Robert Nisbet, Sky—U.S. correspondent; Sally Reardon, APTN editor; and Janet Harris, independent documentary maker. These interviews were conducted between February and April 2011 and via Skype, telephone, in person and via email. Between them the interviewees have variously worked for all major broadcast news organizations based in the UK, including APTN, the BBC, Channel 4, ITN and Sky, and have covered most of the major disasters reported in recent years.

6

Compassion, Nation and Cosmopolitan Imagination

THIS CHAPTER CONTINUES THE DISCUSSION OF MEDIA WITNESSING AND THE MORAL imperative to care from the previous chapter and examines in detail the idea that the mediation of disaster will, in some cases, become a source of cosmopolitan imagination and mobilizer of global compassion. In Chapter 5, the discussion of the injunction to care was based on interviews with journalists, and here we discuss how compassion for others is constituted through media representations. In our analysis, we focus on the news coverage of the Indian Ocean tsunami on Boxing Day morning in 2004 to examine the potential of national and global news representations to cultivate our capacity to feel compassion by nourishing our emotional imagination of distant suffering. The tsunami disaster was extraordinary because of the massive media attention, lasting longer in the headlines worldwide than any other prior disaster (World Disasters Report, 2005). Moreover, it resulted in the best-funded international emergency response in history (Telford, Cosgrove, & Houghton, 2006).

The concept of compassion refers to the moral obligation and sensibility to respond to the suffering of others. Compassion, as several commentators have suggested, is not (only) a "natural" expression of human engagement and care but a socially constructed discourse that imposes meanings on suffering (Ahmed, 2004; Berlant, 2004; Höijer, 2004; Spelman, 1997). However, as a public discourse, compassion has become so naturalized over time that it appears at the very core of the cultural value system of modernity, providing a moral framework for responding to the pain of others (Sznaider, 1998; Tester, 2001). Sociologist of morality Natan Sznaider (1998) crystallized the dictum of Western humanism as follows: "We are compassionate, and if we are not we ought to be" (p. 128).

The world's neglected humanitarian emergencies, together with some success stories of humanitarian response (such as the outpouring of help from around the world after the Asian tsunami in 2004), have made it apparent that there is nothing simple or natural in compassion. After the Pakistani and Indian earthquake in 2005, the Dutch national newspaper *De Volkskrant* printed a thought-provoking headline about the low public response to this disaster compared to the Asian tsunami the year before: "Sorry Pakistan, your disaster just doesn't have it" (Heijmans, 2005). The "it" in the headline refers to the emotional appeal that is needed to excite compassion and form a commitment to public action. In the headline, the appeal appears as an inherent characteristic of a disaster—some just have it and some do not. As argued throughout this book, the emotional definition and meaning of an event are enacted in and through the media. In his book *Media and Morality,* Roger Silverstone (2007) stressed the role of the media by arguing that they are the principal means of constructing our moral life, as the "boundary works" they perform define "the moral space within which the other appears to us, and at the same time invite (claim, constrain) an equivalent moral response from us, the audience, as potential or actual citizen" (p. 7).

The fundamental problem of the politics of compassion in our world today is the excess of suffering, which results in making choices and setting priorities concerning both the representation of suffering in the media and the humanitarian action (Boltanski, 2000, p. 13). In describing the unstable and unpredictable nature of compassion, Keith Tester (2001) wrote the following: "Compassion is like a jack-in-the-box, waiting to spring and create an appearance of activity but kept in the dark all the time no one opens the lid" (p. 65). This raises the questions of when or under which conditions compassion does make an appearance—and a difference to the suffering—and, most importantly, what is the role of the media in promoting a new emotional and political imagination that supports global compassion. As Lilie Chouliaraki (2006,) argued, it is precisely the lack of preexisting global compassion acting on the suffering of others whenever it presents itself that makes the media discourses and practices imperative.

In recent years, media scholars have given increasing attention to the role of the media in forging moral relationships among humans (e.g., Boltanski, 1999; Chouliaraki, 2006; Silverstone, 2007; Tester, 2001). The questions they have asked include the following: How are suffering and compassion constituted through media representations? How and in which contexts may media representations encourage mediated connections of solidarity and responsibility? This debate on the morality of media concerns the idea of cosmopolitanism as a potential consequence of media technologies, discourses and practices (Chouliaraki, 2006; Kyriakidou, 2009; Nash, 2008; Silverstone 2007). Cosmopolitan compassion, in Ulrich Beck's sense, refers to the expansion of our emotional imagination via transnational media—to the greater capacity to put ourselves in the position of others—and an increased awareness of the interconnectivity of human fates in the world of global risks and crises (Beck, 2006). Demands to transform the national orientation of politics and create a new cosmopolitan community lead to questions about the form and quality of media representations that may promote such transformation (see Corpus Ong, 2009). Silverstone (2007), for instance, theorized that "proper distance" in our mediated interrelationships permits us to adopt the position of the other. Chouliaraki (2006, pp. 196–197), as we discussed in Chapter 3, saw cosmopolitan potential in the narratives of "emergency news," in which the "option of pity" is not limited to "our" suffering or to the suffering of people who are like "us" (as, respectively, in "ecstatic news" and "adventure news"), but the viewer is encouraged to take the perspective of the distant other from a different culture. In sum, those media representations that foster new identifications and emotional bonds with distant others have been seen to play an important role in the process of cosmopolitanization (Beck, 2002, 2006; Delanty, 2001; Levy & Sznaider, 2002). In different philosophical theories of compassion, as well as in the academic study of the role of the media in cultivation compassion, the ability to imagine others' suffering is seen as the basis of compassion (Boltanski, 1999; Cohen, 2001; Ignatieff, 1998; Nussbaum, 1996a; Seaton, 2005; Smith, 2002; Spelman, 1997; Sznaider, 1998). What is called for, then, are media representations which foster a new emotional imagination that allows us to see the Others compassionately and, conversely, to see ourselves from the Others' perspective, with an understanding of the extensive material inequality. Moreover, as several scholars have pointed out, if news media as "sentimental educator" are to contribute to the cultivation of a cosmopolitan imagination, audience members need to be able to imagine not only the actual suffering but also the possible moral actions they can take (Boltanski, 2000; Ignatieff, 1998; Moeller, 1999).

As we discussed previously, there is also a strand of negative accounts regarding the difficulty or impossibility of the media to form emotional bonds and commitment between audiences and distant sufferers. Critics have suggested that audiences are emotionally overloaded or indifferent because of the ways in which the news media represent human suffering (e.g., Cohen, 2001; Moeller, 1999). However, we take the position that the news media have a crucial role in the development of "cosmopolitan imagination" and "cosmopolitan empathy" (Beck, 2006, p. 7), understood as an extended capacity to imagine and empathize with the suffering of others beyond the borders of our own kin. Our argument is that the power of the news media can encourage cosmopolitan solidarity and, at the same time, structure our political participation. In this view, the emotion of compassion should be approached as something to be learned as well as capable of change and increase, and the media as the potential means of adjusting and expanding our emotional capacity.

The Cultural Politics of Compassion: Distant and Close Suffering

Recent philosophical, social and cultural theory has highlighted that emotion discourses involve complex relations of power working to either reinforce or challenge them (e.g., Ahmed, 2004; Berlant, 2004; Butler, 2004; Spelman, 1997). As the philosopher Elizabeth Spelman (1997) argued in her book, *Fruits of Sorrow: Framing Our Attention to Suffering,* representing suffering unavoidably means ascribing moral evaluations about the meaning of suffering:

> Invoking compassion is an important means of trying to direct social, political, and economic resources in one's direction....Interpretative battles over the significance of a person's or a group's suffering reflect larger political battles over the right to legislate meaning. *The political stakes in the definition, evaluation, and distribution of compassion are very high.* (p. 89, our emphasis)

In Chapter 4, we argued that emotions are intrinsically implicated in political debates and in the practice of policy making and, therefore, have concrete repercussions in the public sphere. We ought to ask, then, what are the political implications of compassion?

The central figure of the debates surrounding compassion is that of the unknown and distant sufferer. Distance—physical, cultural—creates difficult barriers for the compassionate imagination. Distance renders compassion problematic because it, first, diminishes the possibility of accurate information and creates uncertainty about the individual moral responsibility and about what actions are needed and, second, because media representations of distant suffering are bound to be selective (Boltanski, 1999; Cohen, 2001). Proximity, characteristically relations among family, friends or fellow nationals, has been seen to provide preferred contexts for solidarity, compassion, care and generosity, since it makes it easier to imagine ourselves in the situation of the sufferers (e.g., Slote, 2007). Boltanski (1999) has made a distinction between humanitarian and communitarian frameworks of responses to suffering. The latter links to local interests and preexisting solidarities, which take away or limit the uncertainty about who has the moral responsibility to provide assistance and who is worthy of our care. The former is characterized by lack of ties or precommitment to the sufferer, and the action to relieve suffering is motivated only by the perception that assistance is needed.

Critical views underline the idea that emotion discourses provide the means by which significant differences among people are articulated and domination sustained. First, there is the unequal relationship between the one who suffers and the one who witnesses that suffering. As Lauren Berlant (2004) stated, the term compassion denotes distance and privilege: "The sufferer is *over there*" (p. 4, author's emphasis). Second, and as our discussion of the "calculus of death" and "worthy victims" in previous chapters has shown, compassion involves an inclination to identify with particular suffering, at the expense of others (Berlant, 2004; Butler, 2004; Höijer, 2004). It has long been acknowledged by media scholars that news narratives work as a means to publicly demonstrate whose pain is worthy of our compassion and whose pain is not (e.g., Herman & Chomsky, 1988). Compassion, then, has political implications based on the role of the humanizing or dehumanizing effects of discourse. Besides dividing the world into "us" and "them," compassion may also make it easier to feel superior or hostile toward others. As Martha Nussbaum (2003, p. 13), reflecting on the events of September 11, 2001, wrote, "Compassion for our fellow Americans can all too easily slip over into a desire to make America come out *on top* (author's emphasis) and to subordinate other nations."

Next to the many different ways in which compassion can go wrong, scholars across different disciplines have also been concerned about the political ineffectiveness of compassion to mobilize moral action (Berlant, 2004; Sontag, 2003; Spelman, 1997). A common view is that in order for compassion to be politically effective, it must include an element of outrage, that is to say, understanding and judging of the structural injustices that produce the conditions of suffering (Spelman, 1997). A related view is that compassion, aroused for example by media coverage, remains ineffectual if it is not translated into action (Sontag, 2003). Taking compassionate action is typically contrasted with appropriating the suffering of others to produce an enjoyable feeling in oneself. As Hannah Arendt stated, the emotion of pity (she uses the term *pity* for what we call *compassion*) can be "enjoyed for its own sake," which ultimately works to sustain the suffering (1968, p. 89). By analyzing Harriet Jacobs' *Incidents in the Life of a Slave Girl* (1861), Spelman (1997) showed how Jacobs, while working to arouse compassion among her readers for the experiences of suffering caused by slavery, carefully educated her audience about how to respond to her narrative as she was aware of the possibility that her story could be appropriated for political ends that do not serve the interests of slaves but work to reinforce the domination. Spelman (1997), then, reminded us that we need to be particularly attentive to the ways in which representations of suffering can involve opposing ideological values and reproduce cultural hierarchies. Moreover, she emphasized that when making meaning of suffering, it is important to hear from those who suffer and what it means to them. And as we discussed in Chapter 4, emotional expression can be seen as a form of political agency.

All these problems involved in the politics of compassion have been discussed in the context of news representations. One of the central questions is how the distance between spectators and sufferers encountered in the media can be bridged. The literature offers different answers. For instance, Michael Ignatieff (1998) argued that if news reporting were able to provide more time and in-depth analysis for selected disasters and humanitarian action (in a manner of "documentary features"), it would be more likely to motivate audiences to participate in the relief of suffering. Similarly, Andrew Natsios (1997, p. 125) claimed that if media coverage is "particularly thoughtful and continuous," it can serve as a mechanism for mobilizing the resources needed for relief aid. Depicting sufferers as active agents has been seen as central for creating an emotional commitment from part of the audience (Boltanski, 1999; Chouliaraki, 2006; Moeller, 1999; Tester, 2001). Chouliaraki (2006) identified two dimensions of agency in news reports that shape the ways in which audiences are oriented toward the tragedy of the sufferer: agency refers, first, to how sufferers themselves are depicted and, second, to how the relationship between the sufferers and other actors is depicted. When sufferers are depicted as active agents and other news characters are shown as helping the unfortunate (thus, showing that something can be done), they become more humane for the (Western) audience members who, for that reason, are more likely to become moral agents themselves. However, as Tester (2001) has argued, the focus on the helpers' agency is part of the narrative conventions of disaster reporting that work to "naturalize" suffering since "all the ability to act is identified with the Western relief agencies (who become the subjects who are resolving the problem that the victims cannot sort out for themselves)" (p. 94). Another criticism is that depicting victims as active agents who are able to help themselves can become a mechanism of hiding the structural conditions of the suffering. Thus, representing the suffering at its worst and when sufferers are lacking agency calls for the acknowledgment of the universality of suffering (Cohen, 2001).

Cohen's view on going beyond Western audiences' comfort zone relates to the broader discussion around the importance of images in shaping the impact of crisis events on audiences. Compassion requires imagination and imagination is facilitated by images: images of particular victims and concrete cases of suffering. As Luc Boltanski has noted in his study *Distant Suffering* (1999), a table of statistics on poverty aiming to generalize suffering does not evoke compassion with the fate of the sufferers: "Pity is not inspired by generalities....To arouse pity, suffering and wretched bodies must be conveyed in such a way as to affect the sensibility of those more fortunate"(p. 11). However, as we have discussed before, it is also claimed that images of suffering and death will not sensitize us to suffering but make us "turn away" and create compassion fatigue (Moeller, 1999). This view, as David Campbell (2004) argued, typically confuses state policy and public feelings and conflicts with the reality of successful, image-driven disaster appeals. For instance, Kyriakidou (2009), drawing upon focus group discussions, showed that audience members discuss the disaster events mainly in terms of images of suffering they have witnessed and that it is the "shock of the visuals" that makes them emotionally relate to suffering others.

The appropriation of others' suffering has been approached through the concept of the "voyeur," referring to a spectator who is freed from the moral demands to take action to relieve distant suffering but "can sit back and enjoy the high-adrenaline spectacle unfolding on the screen" (Chouliaraki, 2006, p. 145). In David Morgan's (2002, p. 315) view, the representation of suffering "as a conscious form of 'infotainment'" works to reduce moral sentiments to aesthetic judgments (of the beauty of close-ups or accounts of blown-up and torn-apart people). In this view, fleeting, spectacular images of human suffering work to naturalize and depoliticize suffering by blinding us to its historical and political causes and, in doing so, nurturing a sense of political impotence and voyeurism (Boltanski, 1999; Hammock & Charny, 1996; Ignatieff, 1998; Morgan, 2002). Taking a more optimistic view of the potential of media infotainment to foster global solidarity, Kate Nash (2008) provided a compelling analysis of the role of national and popular media in "creating obligations towards people suffering outside the nation" (p. 168). She discussed the year-long UK media campaign, "Make Poverty History," which was intentionally constructed as highly dramatic—as show business. The campaign was also characterized by an explicit strategy to elicit strong emotions, especially those of pride, joy and righteous anger (rather than compassion, or shame and guilt), in its attempt to create a collective, cosmopolitan "we," who can put pressure from below on politicians and states to create policy in the name of global solidarity and citizenship (by, for example, writing letters and e-mails). Nash concluded that while the campaign was successful in mobilizing public emotion for the suffering of distant people, it failed to achieve "genuine cosmopolitan solidarity," which is more than an identity, as it also requires *"social relationships across differences,* the shared appreciation of material risks and benefits that are unevenly distributed and yet experienced as of common concern to the group" (p. 176, author's emphasis). She noted the following:

> The identity of the global citizen, feeling cosmopolitan solidarity with suffering non-citizens far away *can* be constructed from within the cosmopolitanizing state. What is much more difficult, perhaps impossible, is to construct equitable *social relations* across borders from within the territorially bounded public space of a single state. To do so would surely require not just more equitable structures of global governance, but also genuinely popular transnational media, in which material commonalities and differences are debated from divergent socio-economic perspectives as well as creating and sustaining emotionally charged campaigns like Make Poverty History. The restructuring of global governance and such a transformation in media consumption and production both seem unlikely, however, without a huge transformation in our conception of ourselves as "people" who live in one particular territory and who make up "a (global, political) people." And where is such a transformation to take place except in the popular media? (Nash, 2008, p. 179, author's emphasis)

In sum, on the one hand there is a common understanding that the mediation of suffering can incite compassion and action by creating global solidarity, but on the other hand, media representations are believed to provide occasions for the "commodification of suffering." The contemporary "crisis of pity" (Boltanski, 1999, p. xvi) or "compassion fatigue" (Moeller, 1999) is usually seen to be a result of current market-driven journalistic practices, routine narrative conventions (such as focusing on helpless sufferers and Western heroes) and sensational news values with an emphasis on problems (instead of solutions). How can we overcome the traps of representation? Certainly, it is difficult to theoretically identify which narrative techniques, elements and frameworks in disaster reporting would best elicit appropriate feelings in the audiences. Given the different political possibilities of emotional discourse, the links between media representations and public compassion must be examined case by case. What is required is an empirical and contextual approach that investigates how specific media representations may enable or disable compassion and cosmopolitanism. Certainly, it is too simplistic to assume that media genres, news values or stereotypical news frames determine the emotional responses of audience members (see Wilkinson, 2005, p. 147).

There are only very few empirical studies which examine what kind of emotions are mobilized and how audiences create meanings out of the narratives of suffering. These few studies have nevertheless shown that audiences' emotional responses are necessarily embedded within specific historical, social and cultural contexts and that the responses are far more complex than studies based on theoretical constructions of an audience suggest. Birgitta Höijer's (2004) analysis of Swedish audiences' responses to the media reporting of the Kosovo War, while showing that audience response was conditioned by the cultural construction of the ideal victim, challenged the compassion fatigue thesis by showing how that there are different forms of compassion as well as indifference. Maria Kyriakidou's (2009) study, drawing on focus groups with Greek participants, on the other hand, showed that while the media enable a cosmopolitan imagination, the nation remains a primary framework through which audiences make sense of global disasters and form emotional identifications with distant sufferers. As we argued earlier, the geopolitics of disaster reporting both supports and complicates the nation-centered framework. In the following, this interplay between cosmopolitan and national discourses in disaster reporting is further examined in the context of compassion.

Global and National Context for Compassion

Before addressing the global-national dialectic with respect to the tsunami reporting, we will address the differences of the cultural politics of compassion between national disasters (in Western countries) and distant disasters. We can start with perhaps the obvious fact that unlike many distant emergencies that receive selective media attention, national disasters (such as rail accidents, fires or industrial explosions) are inclusively covered and intensively ritualized today in the national media. While global and national disaster coverage offer similar narratives that articulate and enact compassion, a national discourse of compassion has different political functions as well as moral requirements for the audience. In the context of distant suffering, eliciting compassion through media representations is a crucial means of collecting resources for the relief efforts. When a major accident takes place in London or Helsinki, the audience is not faced with pressure from the national news media to alleviate the situation. As we shall discuss in Chapters 7 and 8, it is the state's responsibility in disaster situations both home and abroad to come to the aid of its citizens and compensate victims. In the past when the government relief and compensation policies were less advanced, disaster coverage also served the concrete and very important purpose of mobilizing resources for the disaster funds that provided compensation to the victims and their families. In the UK, for example, the press reporting of national disasters prior to the 1990s was very much focused on the disaster funds and fundraising events. Until the 1950s, the coverage of major national disasters included daily updated lists of individuals and institutions that had donated money to aid funds, including the amount, no matter how small (Pantti & Wahl-Jorgensen, 2007).

Today, the reporting of national disasters typically mobilizes compassion to manage the anxiety caused by the horror of disaster and to strengthen the emotional bond to the nation. Reporting on national disasters constructs a structure of "compulsory empathy" (see Nguyen, 2009) that is imposed on national audiences as a whole. Audiences are compelled to empathize with the nation since the death and pain of fellow citizens are represented as "our" loss, and the disaster is dramatized as something that speaks about core moral and social values of the nation. The discourse of compassion invites the audience to imagine the nation (rather than others' suffering) as a united community through uplifting stories of heroic rescue workers, common values and common efforts to help (Kitch & Hume, 2008; Pantti & Wahl-Jorgensen, 2007). The main traditional themes of these stories are the restored belief in the goodness of people and the breaking down of boundaries of class, gender or age in the face of disaster. Here is an example from the coverage of the Moorgate tube crash on the London Underground, which occurred on February 28, 1975: "Despite the tragic circumstances, it has restored my faith in the human being. We have had a tremendous amount of calls from people who wanted to offer help, from nursing to mechanical work, and simply making cups of tea" (*The Times,* "Relatives of missing ring up in forlorn hope that Tube crash victims may still be alive," March 3, 1975).

This is not, of course, to say that reporting on national disasters is always consensual and integrative. As emphasized earlier, there are contending emotion discourses involved in the reporting of both national and global crises. The coverage of the tsunami disaster in Finland and Sweden, as we will show in Chapter 8, shows how mediated disasters can be occasions both for global community building and criticizing national political institutions. The tsunami disaster was a global crisis, but it was also a major national disaster for many Western countries (Kivikuru & Nord, 2009), wiping away the usual contrast between the reality of others' suffering and the safety of our lives. About 290,000 people were killed by this rare flood wave, most of them from Indonesia, Sri Lanka, India and Thailand. Among the victims were also 543 Swedes, 539 Germans, 178 Finns and 149 British. As stated earlier, the capacity for identification, or the imaginative power to reconstruct others' conditions, plays a major role in connecting the reader or spectator to the sufferer.

The tsunami disaster was exceptional because the Western audience was not only in the position of a witness and a helper but also of a sufferer. This allowed the Western media to capitalize on cultural proximity and create emotionally arousing and morally engaging stories about our suffering and the suffering of others like us. However, as we will argue, the national resonance also brought those normally experienced as distant others closer, as global commitments and common humanity—shared vulnerability and generosity—were communicated through a national prism. Across the globe, journalists were constructing a global community with rhetoric that fostered equality and solidarity (Pantti, 2010; Robertson, 2008). The following account, published in the *Financial Times,* was representative of many such news reports:

> This has been the world's first truly global disaster, dwarfing any other single event in the emotion and support it has generated and demonstrating graphically that we are indeed one human family, ever more susceptible to common risks and with a shared responsibility to tackle them. (*Financial Times,* "Optimism Rises after the Tsunamis," January 11, 2005)

Before addressing the question about how the tsunami coverage promoted global compassion, we first need to look at how media representations are embroiled in the politics of compassion.

The Asian Tsunami: Cultivating Cosmopolitan Sensibility through the National Prism

Emotions have usually been staged and shared in national settings. Today, the emotional engagement and sharing of emotional experiences incited by transnational media are seen to play a key role in the cosmopolitan world (Beck & Sznaider, 2006). There is, however, a great deal of concern about the capability of the media to expand our emotional imagination and moral action toward suffering beyond the West. Some have singled out the tsunami as an example of mass media-induced global compassion or a "cosmopolitan moment" (Robertson, 2008). However, critical voices have raised serious questions about the global unity created through the transnational mediation of a crisis and, in particular, about the cosmopolitan quality of the tsunami reporting (Chouliaraki, 2008b; Kyriakidou, 2008). Chouliaraki (2008b) pointed to the Eurocentric bias in the construction of moral agency in the transnational tsunami coverage: "The care for the suffering of distant others expands beyond the West only in so far as the West is part of this suffering, both experiencing and witnessing it" (p. 343). Here we can see how the structure of "compulsory empathy"—expressed through disaster narratives that focus on "our" victims and suffering—works also in a negative fashion by providing few other options for empathy and therefore limiting the moral imagination of national audiences (Nguyen, 2009).

Certainly, as we discussed in Chapter 2, the media are routinely accused of interpreting events through a national prism, and the tsunami disaster was no exception in this respect. Indeed, both media organizations and the international aid community have been criticized for working and interpreting the disaster through a national or Western prism. Much of the focus of the Western media was on the national victims and on national generosity as well as the performance of national institutions and international relief agencies. Significantly less attention was paid to losses in tsunami-affected countries and relief efforts and resources provided by local communities and governments (CARMA, 2006; Telford et al., 2006).

The question about the national media's potential to address their audiences as cosmopolitans is a continuation of the debate on the proper form and quality of media representation capable of cultivating cosmopolitan imagination and empathy. Scholars have questioned the assumption that transnational media are actually more likely than national media to promote cosmopolitan sentiments (Pantti, 2010; Hellman & Riegert, 2010; Robertson, 2008). Maria Hellman and Kristina Riegert's (2010) analysis of the coverage of the tsunami disaster by the transnational channel CNN and the Swedish national television channel TV4 is one of a few attempts to empirically compare the role of transnational and national media in producing cosmopolitan empathy and identities. The authors concluded that even if the national news channel framed the disaster as a predominantly national one, it also managed to "bring home" the suffering of local populations through its emotional and personal address. The CNN coverage gave more space to local actors in the affected regions. The authors showed, drawing on both content analysis and focus groups, that CNN failed to create a sense of proximity between audiences and sufferers. This was the result of their distant and factual style of reporting and the clear separation of scenes of suffering "over there" from the viewers' zone of safety, thus providing few possibilities for identification with the victims. The study also raises an interesting question about the national language of the news, which merits further reflection. The Swedish respondents (although fluent in English) felt that English as the news language decreased the sense of proximity and emotional involvement with the mediated suffering. Alexa Robertson's (2008) more extensive comparative study of five national European broadcasters (France, Germany, Sweden, Spain, the UK) and three European channels (BBC World, EuroNews, Deutsche Welle) asked similarly whether global broadcasters are the most liable to foster "cosmopolitan consciousness" (as it is often assumed) or whether news addressing national audiences may also establish cosmopolitan imagery. She concluded that global broadcasters did not contribute more to the mobilization of cosmopolitan imagination than national broadcasters.

The literature on media and compassion centers on television's moral agency. Because of its specific characteristics (visuality, in particular), television has been claimed to have a greater capacity for arousing emotion than print media (Grabe & Bucy, 2009). However, newspapers still have a central role as a news source in some countries (as radio in some), and they are often believed to be more effective than television at providing the detailed information necessary to understand complex policy issues, such as the background to conflict or crisis. We now discuss the coverage of the Asian tsunami in a large-circulation Finnish newspaper.

There is no doubt that Finnish newspapers constructed the tsunami as a Finnish tragedy and that the coverage became a form of nation building. The circle of Finns entitled to victim or mourner status was extended from those personally affected and their families to the friends, neighbors, classmates, people from the same area, people who could have been there, or had been there earlier—all Finns. In the leading national broadsheet, *Helsingin Sanomat,* the grieving subject is indeed the nation, altered by the loss it has suffered: "Finland, perplexed with grief, awakens on Saturday to the year's first morning with a different face" (January 2, 2005). Newspapers paid significantly less attention to directly affected areas and local losses than to Finnish victims. Their suffering was told, for the most part, through mortality figures or through extreme exemplaries: accounts of people who had lost everything. This is one such typical story, originated from Reuters: "Said Taufik, 40, lost nine family members, among them wife, son, father, cousins and grandfather. His house was shattered and there is nothing left from his furniture shop" (*Aamulehti,* January 5, 2005). In contrast, the death of Finnish citizens was presented as a grievous loss, with narratives about individual victims and their families, what their hobbies and interests were or what hopes and dreams they had. The number of stories focusing on affected areas and local people increased slightly after the evacuation of all Finns was completed in the first week of January 2005. The frames through which the stories were told also varied from newspaper to newspaper. While in *Helsingin Sanomat,* about 60% of the articles had a Finnish point of view (focused on Finnish actors and victims); in the largest-circulation tabloid newspaper, *Ilta-Sanomat,* it was almost 80%. The nationalist agenda of tabloid papers is not surprising, but more important than the predominant frame is how the emotional imaginations of readers were cultivated—if they were educated to empathize with others who are not like us. In *Ilta-Sanomat,* the national focus also worked to limit the scope of compassion for Finnish suffering by erasing the suffering of the local people. In the following extract, for example, the journalist reporting from Khao Lak, Thailand, bizarrely turns a Buddhist monk into a voice of Finnish victims:

> After asking where the guests come from, he [the monk] becomes silent for a moment. "A lot of Finns died here," he says. When he hears how many are missing, Chanyoot invites us to a new ceremony. This ceremony he wants to hold only for the Finnish victims. All Thai people on the beach seem to know the tragedy of Finns and other Scandinavians. (January 7, 2005)

The Finnish press was by no means free from public criticism for promoting the culture of taking care of one's own. Newspapers, however, dismissed the criticism by pointing to "universal" news values, such as cultural proximity, and to the news media's "inherent" higher responsibility for national needs and interests. As expressed in the editorial of *Helsingin Sanomat,* "All glory to the global viewpoint but in every country, in the midst of the distress and shock of their own citizens, the mass media's and public officials' duty is to focus on those things" (December 29, 2004). However, the universal values that supposedly overdetermine the reporting have been put in serious doubt by studies that have shown considerable differences in how the tsunami disaster was framed between different national news channels and between national and global channels (Rahkonen & Ahva, 2005; Hellman & Riegert, 2010; Robertson, 2008). For example, German broadsheet newspapers, *Die Welt* and *Frankfurter Allgemeine Zeitung,* did not interpret the disaster through a national prism (Rahkonen & Ahva, 2005). Whereas Finnish and Swedish newspapers focused on national grief, represented by a strong national and emotional imaginary (for example, national flags, religious ceremonies, candles, victims' images from personal albums), German papers aspired to generality by using a distant style (Rahkonen & Ahva, 2005).

The national angle of media coverage is claimed to block the option of compassion for the suffering of people who are not like us (e.g., Chouliaraki, 2006). What we want to argue here is that despite (or because of) the dominant national framing, reporting can actually bring readers closer to distant others and mobilize moral action. In Finnish newspapers, this was done by fleshing out the shared experience of humanity, representing the distant victims as equal moral agents and giving a model of moral behavior. Martha Nussbaum (1996b) has pointed to the false opposition between patriotism and cosmopolitanism. She suggests instead that they are dependent on each other since love for one's own culture enables compassion for other cultures. As such, cosmopolitanism is grounded in particular localities, in that it describes the infiltration of people's everyday local experiences and moral life—worlds with values that orient them toward the distant others (Beck, 2002; Beck & Sznaider, 2006). In the following, we examine how the global and the local intersect in the newspapers by looking at three narratives which, even if they were focused on Finnish actors' feelings or actions, can be seen as facilitating a cosmopolitan imagination. These narratives arise from within a) the stories of Finnish victims' gratitude toward local people, b) the stories of real-world citizens and c) the stories of the "wave of compassion"—the extraordinary willingness of the "ordinary people" to participate in the aid efforts.

In Finnish newspapers, the moral status of beneficent actor was commonly given to the local people who unselfishly and generously helped tourists in the midst of their own suffering. Finnish victims figured prominently in the news and their stories evoked compassion for the local people. Luc Boltanski (2000), referring to Adam Smith's *The Theory of Moral Sentiments* (1759/2002), introduces compassion as a "composite sentiment" (p. 8). Smith imagines a situation in which a sufferer is placed next to a benefactor, with the result that the spectator's sentiments will be a composite of compassion for the beneficent agent and for the sufferer's gratitude:

> When we see one man assisted, protected, relieved by another, our sympathy with the joy of the person who receives the benefit serves only to animate our fellow-feeling with his gratitude towards him who bestows it. When we look upon the person who is the cause of his pleasure with the eyes with which we imagine he must look upon him, his benefactor seems to stand before us in the most engaging and amiable light. (Part II, Sec. I, Ch. I)

Here is one of the many stories of the generosity of local people for which the Finnish tsunami victims displayed gratitude. A seriously injured Finnish woman told her story about the hospital where she was treated in Thailand, remembering local people, and particularly a 10-year-old girl, who helped her:

> 'There was horrible screaming and suffering. There were small children without mothers and fathers. Many were really seriously crushed. "Oh, that one ten-year-old little girl who cleaned sand from my ear," Kivi sighs. "But that [Kivi's own suffering] is nothing in the end as I still have my family and home. Thousands have their whole lives totally shaken," Kivi thinks. (*Aamulehti,* January 3, 2005)

This story was typical, not only in that it acknowledges local people as benefactors with whom the reader can empathize but also in that it shows Finnish victims' awareness of the inequality of suffering, the differences between the experiences of Westerners, who can go back to their lives, and those of local people, who are left with nothing. Such stories were not unique to the Finnish coverage, as they were told in other European countries as well (Robertson, 2008).

The local people also gained the status of moral agents and identifiable subjects of suffering through the stories of ordinary Finns who had already developed cosmopolitan sensibilities. A full-page story about a 29-year-old baggage handler, Jani Holland, at Helsinki International Airport, was especially telling regarding cosmopolitan values. Holland is said to have left his work without taking any luggage with him because he was so affected by the news of the disaster. In Thailand, he reportedly helped the victims and rescuers in the disaster area without an official position. It is not Holland's behavior that the report offered as a model (the dominant message of the overall coverage is that ordinary people can best participate by donating money), but his cosmopolitan disposition and values are defined by the openness to other cultures and willingness to engage with others. The report constructed a stark moral opposition between cosmopolitan and noncosmopolitan dispositions:

> On some beaches, you could see a backhoe or a bulldozer pulling down a house where people have died. A hundred meters away, a fat German orders a piña colada and a massage.…In the evenings, in the bar quarter of Patong Beach, things are much as they always were. Styrox glass-holders keep beers cold, the booze and the parasol cocktails flow, AC/DC hits belt out from the loudspeakers, and tourist men have their hands all over their local girlfriends. One of Holland's acquaintances, working in a bar, is as competent at her job as ever. She carries drinks, arranges chairs and tables, smiles at everyone, cracks jokes. When closing time comes around, the young woman breaks like a twig. She starts crying and throws herself on Holland's shoulder. Her two sisters are dead, one of them in the basement of a shopping mall in which dozens of people drowned. Holland buys two shots of Thai whisky and pours them on the ground in memory of the dead. It is the local custom. (*Helsingin Sanomat,* January 8, 2005)

In contrast to cosmopolitan Holland, who recognizes the suffering of local people and is familiar with their customs, the German sipping a piña colada and the tourist men groping local girls are represented only in terms of egoistical interests and their lack of consideration for the suffering (and bodies) hidden beneath the touristy surface. The narrative fostered a long-distance community of empathy by enabling the readers to identify with Holland and, at the same time, with the Others' loss and vulnerability.

The sense of political impotence is usually seen to be the biggest problem in forming the commitment to alleviate the suffering of distant others. Action that appears to be within reach is crucial to preventing emotions from drifting toward the fictional or remaining in the world of representation (Boltanski, 1999).

According to Boltanski (1999), we can bring compassion into our everyday politics not only by donating money but also by "speaking out in support of the unfortunate" (for example, participating in protests and public opinion polls, pressuring governments to take action; pp. 153, 184–185). The tsunami coverage was exceptional in its emphasis on "ordinary people" as active moral agents who have the capacity to relieve suffering. Compared with routine reporting, they were given far more space in the media than usual (Kivikuru, 2006; Pantti, 2010). Similarly, Robertson (2008) suggested that ordinary people figured prominently in the reports of national broadcast channels in Europe (but had a significantly less important role in all three global channels studied). Finnish newspapers offered a space for citizens' participation in discussing the relief efforts and the politics of compassion. Ethical debates started with a dispute over New Year's fireworks. Setting off fireworks at a time when so many people were suffering and mourning was seen as inappropriate, and people were encouraged to donate the money reserved for fireworks to crisis aid. What followed this particular issue were larger questions of politics of compassion regarding, for example, the hierarchical ranking of sufferers (Should the Finnish victims of the tsunami be in a better position than other victims?) and disasters (Is the tsunami getting too much attention, at the expense of other emergencies?). Citizens also directly proposed different solutions through letters and short message service messages. The solutions ranged from short-term ways to help tsunami victims, such as releasing a stamp with an extra charge to commemorate the Asian flood catastrophe, to adopting a more direct way of helping sufferers by giving a percentage of one's income every month to humanitarian organizations.

One of the predominant stories told during the tsunami coverage was about the unforeseen generosity of Finnish citizens. A half-page report with a typical breaking-records-style headline, "Gains of Finnish collections will rise over 20 million euros" (*Helsingin Sanomat,* January 9, 2005), was especially telling when it comes to offering moral and sentimental education to the reader. It opened with a story of Tuulikki Hynynen, manager of the Red Cross division in the small municipality of Siilinjärvi in Eastern Finland, collecting money for the tsunami disaster:

> "Is this going there . . ." asks a middle-aged woman before she drops a fiver into the box.
>
> "Yes, to the disaster area of Asia. Thank you". . . .
>
> People who donate money make eye contact with the collector already from a distance. People who are going to pass the collection don't look into her eyes.
>
> "I have already participated," many shout. . . .

> "One has to give something every time," says a smiling woman and coins clack into the box. Children get a FRC sticker on the back of their hands as a thank you.

The story ends when Hynynen is back at the Red Cross office. She is speculating as to how much money she has collected, although she knows that the exact amount will not be known until she brings the box to the bank, "where the seal will be broken and the money transferred to the RC's catastrophe fund."

The donation stories communicated three dominant messages that aimed to mobilize compassion. First, these stories, as well as record-breaking news about aid collections (always including mobilization information details such as bank account numbers of aid organizations), showed that donating money, not teddy bears or old clothes, is the right way to respond. Focusing on reports of volunteers collecting money, and people giving on the streets, the audience was provided with a straightforward model of moral action (and thereby empowered). The moral importance of ordinary citizens was further emphasized by contrasting their generosity with the "shamefully" stingy pledge of the Finnish government (which was raised significantly after the public criticism) or the selfish strategic political agenda of other governments, in particular that of the United States, which stood to gain from the global generosity contest. Second, donation story lines strongly communicated the reliability of relief organizations like the Red Cross. In the newspapers, there was a consensus about the legitimacy and competence of aid organizations, and the journalists did not feel compelled (this time) to stress the problems relief organizations face when delivering aid to the disaster area. Third, the reporting suggested that taking part in relief aid translates into rewarding feelings. People participating in collective action were depicted as entitled to feel good and proud of themselves. Additionally, the coverage moralized those who did not participate with opposite feelings: shame and guilt. Shame and guilt, along with compassion, are "cosmopolitan emotions," since collective responsibility and solidarity depend also on feelings of shame and guilt when little is done to relieve suffering (Linklater, 2007). This time, then, the national frame was not as inclusive, because it constructed a moral border between those who feel responsible for others and are able to look directly into their eyes, and those who cannot.

Conclusion

We have argued here that media representations of suffering are crucial in extending the emotional imagination to include people who stand outside the traditional boundaries of moral concern and in fostering a sentimental education that allows extending our care and moral obligations. We have taken the coverage of the tsunami disaster as an example of a disaster in which the media occupied a central position as an organizer of both moral emotions and humanitarian aid. This time, it successfully participated in facilitating the imagining of others' suffering and prompted the unpredictable emotion of compassion. The question is whether the tsunami coverage has brought about a more permanent change in the emotion discourses of disaster coverage. There is only anecdotal evidence to support the claim about its longer-term moral power and public consequences. It seems that the narrative about the generosity of local people and gratitude of the afflicted Western victims used in the national media was particularly compelling. In Finland, it became a typical reference when the ash cloud from Iceland's Eyjafjallajökull volcano stranded international passengers at the Helsinki airport in April 2010. In the public discussion (in letters to the editors and on Internet discussion sites), the unwillingness of the Finnish authorities to do anything to help stranded passengers was heavily criticized and commonly compared to the generosity of the Thai people during the tsunami. The argument was that Finns should similarly strive to display compassion and show hospitality to the strangers in need.

The tsunami became heavily ritualized by the news media and was subject to the discourses of national identity, globalization and cosmopolitan citizenship. In this chapter we have dealt with the question of the role of national news media in establishing cosmopolitan imagery and inciting compassion for distant others. What emerged in the Finnish tsunami coverage could be defined as a moment of "internal globalization," a transformation of the public sphere at the national level as global concern became part of local experiences and moral life worlds (Beck, 2002). Representing suffering with a strong national reference, with local people in the afflicted areas featuring as moral agents in the role of collaborators, resulted in a "wave of compassion." News about the relief response, operating on a sphere of consensus, forcefully created an active public and invited the readers as citizens to participate in public action. This was done, in one way, by giving citizens an opportunity to participate in the moral discussion about the meaning of the tsunami. The news coverage offered plenty of models for feeling right and solutions for practical action. It is fair to say that Finnish newspapers, in particular the tabloid papers, overstressed the Finnish point of view, but in the process of narrating and portraying "our" suffering, it also succeeded in producing proximity with distant others. Other studies on national tsunami coverage have come to the same conclusion (Hellman & Riegert, 2010; Robertson, 2008). This supports the claim, underpinned by the geopolitics of disaster coverage, that strong identifications and feelings toward others are only produced when distant events have a local resonance

(Levy & Sznaider, 2002), and it gives more reason to reflect upon the role of global news channels in mobilizing the cosmopolitan imagination.

For the journalists working in the disaster area, as well as on the home front, the need to make sense of the terrible suffering was obvious. As we have seen, the dominant narrative emerging from the tsunami reporting highlighted the common humanity based on *shared vulnerability* and suffering on the one hand, and on compassion as a *natural* feeling that brings people together on the other. Tsunami reporting, therefore, helped to shape the creation of a "cosmopolitan outlook" (Beck, 2006), forced into being by representations of shared global threats (see also Nussbaum, 2001: 320; Frosh & Pinchevski, 2009: 301–302). Stressing our common vulnerability can be a powerful tool for cosmopolitan education and encouraging cosmopolitan empathy that extends beyond the realm of the nation state.

7

Disaster Citizenship and the Assumption of State Responsibility

THE PRECEDING TWO CHAPTERS HAVE DEMONSTRATED THAT THE WITNESSING OF disasters places journalists in the difficult role of weighing between the professional calculus of death and the injunction to care. They argued that even if in negotiating these complexities, concerns for the "home nation" remain at the forefront, a space remains for the cultivation of cosmopolitan sensibilities based on the forging of emotional connection. This chapter builds on the understanding of the ideological inflections of witnessing and the injunction to care by examining how they are shaped by "our" constructions of the suffering of others. The Burma cyclone and China earthquake of 2008 and the 2010 Haiti earthquake offer useful comparative cases of how discourses around disaster coverage are politicized in this way. The chapter examines coverage of these three disasters in major international newspapers in the two weeks following the events as well as in selected broadcast and online reporting. The disasters occurred in three very different national contexts, which are nevertheless all viewed in Western liberal discourses as embodiments of problematic governance regimes: China, ruled by the Communist Party, is, as the world's most populous nation and now the second-largest economy, emerging as a key player on the world stage. Burma is known for its long-standing repressive military regime, which has ruled the impoverished and isolated nation with an iron fist since 1962, suppressing dissent and opposition with unrestrained violence. Finally, Haiti is one of the world's poorest nations, marked by a history of colonialism and brutal dictatorships. The country has been effectively ruled by the UN since 2006, and according to dominant discourses, its continued desperate state underscores the failure of international intervention.

This chapter investigates discourses on the role of the state on the one hand, and the international community on the other, in the coverage of these three disasters, highlighting the specific manifestations of what we refer to as "disaster citizenship." The disasters brought about what we have previously described as moments of heightened reflexivity: They put the spotlight on the regimes in question and, in doing so, consistently raised questions of political import, albeit in very different ways in each of the three cases.

As scholars have argued (e.g., Sorenson, 1991), disasters that take place in non-Western, impoverished and postcolonial nations allow for the dramatization of representations of "the Other" in relation to "Us." Earlier in the book, we discussed how distinctions between "Us" and "Them" derived from the geopolitics of disaster coverage shape, among other things, the ways in which emotional responses are prefigured and elicited in disaster coverage. These distinctions inform decisions about "worthy" and "unworthy" victims and are used as means for doling out or withholding compassion and empathy. The discourses of disaster coverage (as, indeed, those of other forms of journalism) are profoundly ideological, and the invocation of the nation state is crucial in producing difference, even if "Us" and "Them" distinctions are also, as we have argued, made on the basis of other forms of difference, including race, class and gender, which interplay in complex ways with notions of citizenship.

Yet we suggest that "Us" and "Them" distinctions also derive from a more subtle set of normative assessments around the nature of governance and citizenship in particular countries. As such, the geopolitics of disaster coverage is complicated by modes of governance. In particular, non-Western, nondemocratic nations tend to be constructed as "Other" by Western media coverage of disaster, and their citizens, aside from being victims of disaster, are also constructed as the victims of repressive regimes and/or poor governance. In his classic analysis of media coverage of the war in Vietnam, *The Uncensored War,* Daniel Hallin (1986) described journalists' world (and hence, their professional practices) as "divided into three regions, each of which is governed by different journalistic standards" (p. 116). First of all, what Hallin refers to as the sphere of consensus "encompasses those social objects not regarded by journalists and most of the society as controversial. Within this region journalists do not feel compelled either to present opposing views or to remain disinterested observers. On the contrary, the journalist's role is to serve as an advocate or celebrant of consensus values" (pp. 116–117). Bounding the sphere of consensus is the sphere of legitimate controversy, where norms of objectivity and balance apply and journalists are obligated to represent different points of view. Beyond the sphere of controversy "lies the Sphere of Deviance, the realm of those political actors and views which journalists and the political mainstream reject as unworthy of being heard" (p. 117).

Here, we are especially interested in the operations of the sphere of consensus around disaster coverage. We suggest that it includes (as do probably other forms of international news) what are viewed as uncontestable and commonsense assumptions, expressed through a series of politicized metanarratives. These metanarratives revolve around normative principles relating to the value of democratic governance principles, freedom of speech and suppression, the history of colonialism (in the case of Haiti), the benefits of globalization (in the case of China) and the dangers of authoritarian regimes (in the case of Burma). As metanarratives, they reveal as much about "Us"—or the ideological presuppositions of Western media—as they do about "Them." As we shall see, they are underwritten by what we might call "the assumption of state responsibility": As Naomi Klein observed (2007), the common assumption is that the state will do whatever it can to come to the aid of the victims of disaster. The role of the state as a savior is taken for granted in the case of disasters, and the idea that it will play a central role in the relief efforts is unquestioned. Any shortcoming in this regard is viewed as an abdication of one of the most fundamental duties of the state, that of protecting the lives of its citizens/subjects. This is precisely why the Bush government's handling of the aftermath of Hurricane Katrina was, in the eyes of many observers, so shockingly inadequate: In the world's largest economy, supposedly the epitome of liberal democratic governance, thousands of citizens were abandoned to their fates because the state responded so poorly or,at the very least, unevenly. The result was that the most needy victims were the hardest hit. Katrina dramatized how, although the initial horror of disaster strikes all victims with equal force, the subsequent ability of victims to survive and build up a new life often depends on their resources and the ability and willingness of the state to redress their suffering.

The assumption of state responsibility has profound consequences for the forms of cosmopolitan imagination made possible by disaster coverage. Proponents of cosmopolitan citizenship suggest that a new era of globalization has ushered in a political reality where "few decisions made in one state are autonomous from those made in others" (Archibugi, 1998, p. 204). On this basis, they envision "expanding cosmopolitan democracy and global governance, in which for the first time there is the possibility of global issues being addressed on the basis of new forms of democracy, derived from the universal rights of global citizens" (Chandler, 2003, p. 332). Cosmopolitan citizenship is thus premised on a higher level of public accountability, which takes into account not merely the individual rights of a nation state's citizens but the universal rights of citizens of a global society (Chandler, 2003). More broadly, cosmopolitan citizenship, then, is anchored in the global assertion of the values and practices of democracy (Archibugi, 1998). As such, the debate over cosmopolitan citizenship exposes a fundamental challenge to deeply engrained understandings of national sovereignty:

> The two opposite poles of the spectrum are evident. On the one hand, there stands the principle of sovereignty with its many corollaries...on the other, the notion that fundamental human rights should be respected. While the first principle is the most obvious expression and ultimate guarantee of a horizontally-organized community of equal and independent states, the second view represents the emergence of values and interests...which deeply [cut] across traditional precepts of state sovereignty and non-interference in the internal affairs of other states. (Bianchi, 1999, p. 260)

The model of cosmopolitan citizenship clearly raises difficult questions not just about state sovereignty but also about the universalization of Western values, while recognizing the interconnectedness of the globalized world and the opportunities and responsibilities that come with it. As Silverstone (2007) argued:

> It also includes the troubling and dangerous elision between the taking of global responsibility, humanitarian intervention, and what might otherwise be considered...as colonialism. Cosmopolitan realism, in the acknowledgement that there are no separate worlds, represents a kind of contextual universalism, one which suggests that non-intervention in the crisis of the other is no longer possible because we are, in this new global era, intimately connected as never before. (p. 17)

We have previously discussed how disasters have been seen as opportunities for such cosmopolitan citizenship to emerge—an opening for "cosmopolitan moments" (Beck, 2009) that bring to the fore questions of risks, rights and obligations extending beyond the realm of the nation state, connecting all of us as members of a global community. However, when cosmopolitan citizenship is conceived in this way—as emerging in response to threats that do not obey national borders—it is based on rather a bleak understanding of what brings about the conditions for global solidarity—major disasters which dramatize individual and collective risk. Such an understanding, grounded in what we refer to here as disaster citizenship, is radically at odds with other accounts of cosmopolitanism, such as those spurred on by the explosion of mobility brought about by globalization and the subsequent rise of curious, fearless, open-minded tourists exploring the wonders of the world (e.g., Szerszynski & Urry, 2006). The imagined "new cosmopolitan is assumed to be free from the tying and oppressive loyalties of the singular community. In the ideal world such a figure is mobile, flexible, open to difference and differences" (Silverstone, 2007, pp. 11–12). Disaster citizenship, by contrast, is elicited on the basis of catastrophic and destructive events with profoundly negative consequences for its victims and, by virtue of globalization, for the world as a whole. It is inherently top-down insofar as it is hailed through mediatized representations of distant suffering and the subsequent reactions of state and nongovernmental actors. It is also profoundly reactive, emerging not spontaneously and organically but rather in response to the occurrence of disasters. The possible reactions—the horizon of action encouraged and enabled by disaster citizenship—remain limited by conventional strictures of national citizenship, to the expression of consent or dissent in public opinion polls, letters to the editor and other forums for public debate, and the donation of money and material goods to assist in the relief efforts. It is, in practical terms, a relatively "thin" understanding of citizenship that constructs the citizen as a consumer. Finally, as we have suggested elsewhere in this book and further discuss in the remaining chapters, disaster citizenship unfolds within the bounds of the nation state and the scope of sovereignty, and any cosmopolitan sensibilities emerge from within this context. This chapter suggests that any action on the part of the nation state and the international community which exceeds its customary range appears to require careful and nuanced justification, differentiating between disasters and their consequences in explicitly political terms, premised on the universal value of democracy and the conceptions of citizens' rights, which go along with it. The assumption of state responsibility is central to such justification, but precisely because it remains moored within conventional understandings of national sovereignty, disaster citizenship, however conceived, falls short of an unfettered cosmopolitanism. Nonetheless, it does construct spectators who can care for the plight of distant and different others.

Certainly, the initial tales of human suffering caused by disaster are eerily similar from one event to another, following the template we have discussed elsewhere (Pantti & Wahl-Jorgensen, 2007), sharing information about the horrific scale of the disaster and circulating images of devastated and shocked victims and destroyed buildings. However, in the cases of Burma, China and Haiti, what we have described as the discourse of horror was, even at the earliest stage, accompanied by coverage of the political elements of the disaster.

The coverage highlighted how the scale and quantity of suffering are not evenly distributed but vary according to a complex set of interacting factors. These include the scale of the disaster, the resources and wealth of the country, its political importance, the nature of the political regime and the ability and will of the state to come to the aid of its citizens. As a result, in all three cases, the disaster coverage became a story about the successes and failures of state action. At the same time, the assumption of state responsibility also coexists in a complex relationship with consensual understandings of the responsibilities of the international community and, in particular, aid organizations. Generally speaking, in the initial stages of disaster coverage, humanitarian aid, whether coming from state actors or NGOs, is viewed as a relatively unproblematic and benevolent restorative force, the need of which stands in an inversely proportional relationship to the state's ability to help its citizens. The assumption common to such coverage is that aid organizations promote "the ideals of global humanitarianism" (Cottle & Nolan, 2007, p. 862), both by calling attention to the suffering of deserving others and providing humanitarian assistance to them. Media play a key role as a bridge between the public and aid organizations precisely because their work takes place within the sphere of consensus (Thomas, 2011). When disasters occur, the media will, at least initially and temporarily, dispense with conventions of objective reporting and rally around the relief effort as part of the "injunction to care" discussed in Chapter 5. This practice covers over what some critics have seen as a problematic interventionist and often outright political project of humanitarianism which frequently works to discursively and practically undermine the sovereignty of nations through the imposition of "international therapeutic governance" (Pupavac, 2004) justified by the state of emergency (Calhoun, 2008).

Because of the initial discursive consensus around the benign role of the international community, including governmental and nongovernmental actors, states are expected to welcome aid with open arms and facilitate its distribution, and any action contrary to this is viewed with outraged condemnation. This was very much the template followed in the disasters discussed in this chapter. To be more specific, the coverage of the Burma cyclone offered an opportunity for international media coverage that questioned the actions of the country's authoritarian military regime. The regime's failure to respond to the unfolding humanitarian crisis and the resulting anger of the populace were reported extensively, eliciting compassion for the cyclone's victims as well as, more broadly, raising awareness about the plight of Burma's citizens. By contrast, the speedy and efficient response of the Chinese government to the earthquake provided a narrative which emphasized the perceived new openness of the regime and was particularly interesting in the context of the timing: The earthquake occurred in the run-up to the 2008 Olympic Games in Beijing, which placed China's record on human rights and democracy firmly in the international spotlight. Finally, the Haiti earthquake was framed with a focus on the underlying causes of the substantial death toll, including the history of colonialism, the failed state, poverty and poor construction standards. In all three cases, the politicized nature of coverage and criticism of governance regimes reminded us of the tenuous line between "man-made" and "natural" disasters.

The Horror and Politics of Disaster Coverage

Cyclone Nargis struck Burma on May 2, 2008, causing the worst natural disaster in the recorded history of the country. The reported casualty figures were just over 138,000, but the actual death toll is believed to have been much higher, later estimated at anywhere between 200,000 and 1.5 million, and the disaster also left millions of Burmese people homeless. Eight days later, on May 10, 2008, an earthquake measuring 7.8 on the Richter scale hit Sichuan in China. The earthquake killed at least 68,000 people, and more than 4.5 million people lost their homes. Early reporting in both cases shared information about the immediate details of the disaster but also immediately introduced a political element. The Burmese cyclone was instantly placed in the context of a major unfolding domestic political story, that of an upcoming referendum on a new constitution seen as favourable to the ruling military regime. A story from *The Australian,* published on May 5, 2008, immediately after the earthquake struck, was exemplary of this early stage of coverage:

> RANGOON: Burma's military junta declared disaster areas in five states yesterday after tropical cyclone Nargis pounded the Irrawaddy delta, killing at least 240 people.
>
> Nargis, which was packing 190km/h winds when it hit on Saturday, left the streets of the main city, Rangoon, littered with debris from fallen trees and battered buildings....
>
> It remains to be seen what impact the storm will have on the referendum on an army-drafted constitution scheduled for May 10.
>
> The charter is part of a "road map to democracy" meant to culminate in multi-party elections in 2010 and end nearly five decades of military rule. Critics say it gives the army too much control. ("Burma chaos as cyclone kills 350," May 5, 2008).

This report, drawing primarily on official Burmese sources, immediately called attention to the upcoming election and placed the disaster in the context of a history of "nearly five decades of military rule." This political context was to become the dominant frame through which the disaster was subsequently interpreted by international media. In the case of the Sichuan earthquake, the initial reports emphasized the decisive action taken by the Chinese government:

> A massive death toll is feared after a powerful earthquake struck densely populated areas of south-western China late yesterday.
>
> The quake, measuring 7.8 on the Richter scale, flattened schools and homes in several cities, with up to 5000 people reported killed in one county of Sichuan province alone.
>
> A further 10,000 people in the county, Beichuan, were also feared injured, a report from the state-run Xinhua news agency said.
>
> Chinese Premier Wen Jiabao called it a "major disaster" and was reported last night to have rushed to the area, while President Hu Jintao ordered an "all-out" rescue effort. (*The Age,* "Huge toll feared in Chinese quake," May 13, 2008)

At the same time, particularly poignant coverage told of collapsed schools burying hundreds of children in rubble. This story was to repeat itself in different forms over the next few weeks, as news emerged of hundreds of school collapses across the area affected by the earthquake:

> MORE than 900 students are entombed in the ruins of a collapsed school building after a massive earthquake in south-west China yesterday.
>
> In harrowing scenes, rescue workers struggled amid twisted concrete and metal to free trapped teenagers from what remained of the three-storey Juyuan Middle School in Dujiangyan city. (*The Mirror,* "Buried alive," May 13, 2008)

As we shall see, the widespread occurrence of school collapses following the earthquake became a political question over time, but at this early stage of coverage, the media focus was on documenting and describing the horror caused by the devastation.

On January 11, 2010, an earthquake measuring 7.3 on the Richter scale struck Haiti, one of the world's poorest countries. It took considerable time for reliable news of the disaster to reach the outside world, as communications inside the country broke down. The first images of the disaster circulated on social networking site Twitter. When details emerged, it was clear that the disaster was devastating: Early estimates suggested that at least 50,000 had perished and 3 million citizens were likely to have lost their homes—the casualty figures were later significantly upwardly revised, and it is now estimated that up to 316,000 people lost their lives in the catastrophic disaster and its aftermath. An article in *The Wall Street Journal* captured the prevailing tone of early coverage:

> Cries from victims entombed beneath concrete debris pierced the air of seemingly every street in this crowded capital Wednesday, where shocked residents carried the injured and the dead a day after the nation was hit by a quake that some estimate has killed more than 100,000 people.

> Haitians tried digging through rubble with their bare hands to rescue people trapped after the biggest earthquake to hit the impoverished Caribbean nation in two centuries. Thousands of buildings from shanties to the presidential palace were destroyed, streets were blocked by debris and telephone service was knocked out. Countries around the world, meanwhile, scrambled to send in help. ("Haiti despairs as quake deaths mount," January 14, 2010)

Though this report focused primarily on outlining the scale of the disaster, it also emphasized the "impoverished" state of Haiti and the immediate efforts of the international community to aid the beleaguered nation, thus setting the stage for the coverage that was to come. Underpinning all of this coverage was what we refer to as the assumption of state responsibility: the consensus notion, discussed earlier, that the nation state is responsible for coming to the assistance of its citizens and that governments can and should, therefore, be assessed on the basis of whether they are acting on this responsibility. This, in turn, has consequences for the specific forms of disaster citizenship and cosmopolitan solidarity elicited by disaster coverage, insofar as they are always already grounded in the context of the nation state. In the stories examined here, the assumption of state responsibility is closely tied to the injunction to care: Through the description of the horror of the disaster, the journalist bears witness to the scale of the damage done and the extent to which the state has come to the aid of the victims.

The Assumption of State Responsibility: A Tale of Three Disasters

In the three disasters, the assumption of state responsibility was articulated in very different ways and consequently gave rise to different forms of disaster citizenship. In the case of Cyclone Nargis in Burma, the failure of the dictatorship to come to the aid of the disaster victims was widely condemned and raised important questions about the responsibility of the international community. Following the disaster, it quickly became clear that far from coordinating the relief effort, Burma's ruling elite was actually working *against* it. Early on, the government issued a call for assistance, which was received with caution by international media. A *Belfast Telegraph* article was representative of such coverage, which questioned the sincerity of the Burmese regime by drawing on the critical voices of campaigners abroad. The article ultimately implied that the generals, in their calls for assistance, were motivated more by fear of protest and instability than by concern for the well-being of Burmese citizens:

> The secretive military junta that has ruled the impoverished nation for two decades took the unprecedented step yesterday of issuing an urgent appeal for international help. . . .
>
> Campaigners have said the junta failed to issue adequate warnings that the storm which had been building for several days off the Burmese coastline was about to strike. There were also complaints that the 400,000-strong military was only busy clearing streets where the ruling elite lived and leaving other residents to fend for themselves. "The regime failed to warn people and are failing to help them now," said Mark Farmaner, of the Burma Campaign UK. "One of the few things that may motivate them to allow aid in is the fear of another uprising. People are asking why they can mobilise the police and army to attack democracy protests but do nothing now." ("Cyclone claims 15,000 lives as Burma makes an unprecedented plea for international help," May 6, 2008)

This report, along with many other early stories, highlighted the fact that the regime had failed to warn the Burmese population of the coming storm and hence endangered its citizens. Following on from this early questioning of the sincerity of the generals' call for aid, news reports, editorials and commentaries published in the days after the cyclone struck also increasingly called attention to the failure of the regime to respond to the disaster. Despite the generals' calls for aid, news reports published only a few days after the disaster struck discussed the increasing worry of humanitarian aid organizations and the international community about their inability to provide assistance and the danger of further casualties resulting from the lack of such assistance. As these concerns grew, some began to suggest that the international community should use the UN principle of "the responsibility to protect" to justify humanitarian intervention. An article in Canada's *Globe and Mail* reflected this unfolding debate:

> France, Britain and Germany called yesterday for the world to deliver aid to cyclone victims in Myanmar if necessary without the military junta's permission, France's junior minister for human rights said.
>
> "We have called for the 'responsibility to protect' to be applied in the case of Burma," Rama Yade told reporters as European Union development ministers met to discuss emergency aid for Myanmar, formerly known as Burma. ("Europeans call for forced intervention," May 14, 2008)

Backing up the application of the hitherto obscure and contested principle, an opinion piece in the British *Guardian* made the case for intervention by the international community through the implication that the Burmese government was responsible for the significant loss of life in the aftermath of the disaster:

> It is probably too late to save the first wave of victims in the Irrawaddy delta, but it is not too late to create structures that give teeth to the UN principle of a "responsibility

> to protect"—a principle designed to be invoked for genocides, not cyclones and earthquakes. (*The Guardian,* "No quick fixes for Burma's disaster relief," May 14, 2008)

Here, precisely because of the relative isolation of the Burmese regime and the perception of its blatant disregard for the lives of its citizens, the disaster became a focal point of global political debate centered on the assumption of state responsibility. Intervention was thus justified in commonsense terms of the failure of the state to meet this basic obligation. In this case, the failure was not described as caused by a lack of resources or ability but rather as a deliberate, malicious and self-serving move reflecting the skewed priorities of an inhumane dictatorship. Several stories described how food aid, once it was allowed into the country, was hoarded and sold by the regime while its citizens were starving. For instance, an article in the British newspaper, the *Daily Telegraph,* quoted Burmese volunteers who were making these accusations:

> OFFICIALS in Burma's cyclone-hit Irrawaddy delta area are appropriating emergency aid supplies and selling them in local markets, it was claimed yesterday.
>
> Burmese volunteers who are operating their own private aid missions have said that they are having to hide from local apparatchiks in order to prevent them commandeering their supplies.
>
> The *Daily Telegraph* learnt of the alleged scam from a Burmese businessman from Rangoon who was leading one mission distributing rice, biscuits and clothing. "If they see our relief supplies, they will come over and say 'don't worry, give that to us, we will distribute it for you,'" he said. "But we know that for every 10 sacks of rice we give them, only four will reach the people."
>
> "The other six will end up being sold by that official on a market in some local town. Rice prices are very high right now and that official will then make a good profit." ("Burmese officials 'are seizing emergency aid and selling it for profit,'" May 13, 2008)

The failure of the Burmese regime to come to the aid of its citizens was frequently discussed in the context of governance, as in another *Guardian* leader, which was reprinted in newspapers around the world:

> "No famine has ever taken place in the history of the world in a functioning democracy," Amartya Sen once wrote. This, argued the Nobel-prize winning economist, was because democratically elected governments "have to win elections and face public criticism." Mr Sen's words have a grim resonance as Burma faces a humanitarian disaster exacerbated by its lack of openness. ("Disaster strikes," May 6, 2008)

This editorial, along with several other articles, opinion pieces and letters to the editor, emphasized the fine line between "man-made" and "natural" disasters and the ways in which such distinctions cover over larger political contexts—in this case, how the inaction of the Burmese dictatorship made worse the already devastating damage caused by the cyclone. For example, an *Irish Times* article quoted the Irish minister for foreign affairs, Micheál Martin, in discussing the "responsibility to protect" principle: "The major challenge to an effective response to this crisis is man-made. The challenge is the reluctance of the government of Burma to accept international humanitarian assistance" ("Martin is critical of Burmese regime," May 16, 2008).

Such coverage meaningfully problematized the political contexts of the disaster and their consequences for victims—a theme which was also central to the coverage of the Haiti earthquake. In that case, however, the reasons why the disaster was presented as man-made were more complex: In Haiti, it was not the (in)actions of a brutal dictatorship which were to blame. Instead, the difficulties of the relief effort in Haiti were narrated in terms of the long-term trajectory of the failed nation state which, in turn, meant that the international community and the governments of wealthy nation states had to step in where the government could not, justifying disaster citizenship in terms of the breakdown of conventional nation-state-based governance. Even at the earliest stages, and despite the urgency of disseminating basic information about the unfolding events, many stories also focused on contextualizing the disaster, describing the underlying and structural reasons why it hit the country so severely. Such coverage included the discussion of Haiti's political history, dominated by the violent and repressive dictatorships of Papa and Baby Doc between 1957 and 1986, of poverty, environmental degradation and poor infrastructure and construction. This coverage both provided a critique of global inequalities and the distinction between "man-made" and "natural" disasters, which was also problematized in the Burmese case. Seumas Milne, a commentator for the British *Guardian* newspaper, exemplified this approach in a column published on January 21, 2010:

> . . . while last week's earthquake was a natural disaster, the scale of the human catastrophe it has unleashed is man-made.
>
> It is uncontested that poverty is the main cause of the horrific death toll: the product of teeming shacks and the absence of health and public infrastructure. But Haiti's poverty is treated as some baffling quirk of history or culture, when in reality it is the direct consequence of a uniquely brutal relationship with the outside world—notably the US, France and Britain—stretching back centuries. ("Haiti's suffering is the result of calculated impoverishment")

A BBC online news story focused on examining the poor construction work of most housing stock in the country and opened as follows: "Experts say it is no surprise that shoddy construction contributed to the level of destruction in Haiti following Tuesday's earthquake. But the scale of the disaster has shed new light on the problem in the impoverished Caribbean nation" ("Haiti devastation exposes shoddy construction," January 15, 2010). Based on interviews with architects and other construction industry experts, the article discussed the poor quality of concrete and the fact that most buildings were constructed using masonry, which is particularly inadequate in earthquakes. Michael Deibert, described as a "journalist and Haiti expert," suggested that in fact, deforestation of rural areas in the country was the most significant underlying structural problem contributing to the scale of the disaster: It meant that the quality of the soil was too poor for subsistence agriculture and had resulted in the large-scale migration of people from the countryside to the overcrowded capital of Port-au-Prince, only 10 kilometers from the epicenter of the earthquake.

These discourses focused on structural causes of the devastation caused by the earthquake, establishing the relative hopelessness of the situation in Haiti due to a troubled and traumatic history and thus justifying intervention. However, other coverage placed this in the context of how Haiti had been shaped by global forces. Overall, the coverage was generally careful to avoid the essentialist Othering that has, as discussed earlier in the book, long been a pattern of media coverage (Sorenson, 1991). Through such coverage, the "natives" are constructed as being responsible for the disaster that has befallen them, ignoring complex histories of colonialism and the tangled webs of global capital. Instead, coverage of the Haiti earthquake often discussed structural causes, including the role of the international community and the ideology of neoliberalism. For example, in an analysis of the work required to rebuild Haiti, the British newspaper, *The Guardian,* featured a column by its environmental editor, John Vidal. In it, he articulated a critique of the political economy of foreign aid as partially responsible for the dire straits of the country, echoing the comments of critics regarding what are often ultimately destructive practices perpetrated in the name of humanitarianism (see Calhoun, 2008):

> Reconstructing Haiti is the ultimate challenge to an international community which has failed over decades to lift the island state out of poverty, corruption and violence.
>
> In the past 10 years more than $4bn has gone to rebuild communities and infrastructure devastated by hurricanes, floods and landslides, but mismanagement, lack of coordination and attempts by global institutions to use Haiti as a neo-liberal economic testbed are widely believed to have frustrated all efforts. A foreign debt of $1.5bn has weighed down the economy. Last year, the government paid $79m to service debt, but received under half that to support schools, health and transport. The danger is that money expected to pour in to rebuild

> the shattered state will be again misappropriated by an elite, or serve to further undermine the government. ("Haiti: Learning from past mistakes is the first priority," January 15, 2010)

Such critical discourses, given prominence in mainstream mass media, fed into a metanarrative about the limits of neoliberalism and the damage wrought by post-colonialism and reminded audiences, once again, of the "man-made" nature of the disaster. Thus, rather than supporting the status quo of global power relations and economic practice, they challenged the presumptions underlying foreign aid and neoliberal regulatory regimes. As such, these critical discourses on Haiti echoed the case made by Naomi Klein in *The Shock Doctrine* (2007), where she outlined how a wave of neoliberalism has swept across disaster-struck nations and ultimately exacerbated the plight of already impoverished victims.

In making this case, such stories differed from responses to the Burma disaster: They focused on the responsibility of the international and global community, rather than primarily assessing a national regime. This may in part have been due to the widespread recognition of, and consensus on, the absence of a functional state in Haiti. As much coverage pointed out, the country had effectively been under the governance of the international community, in the form of the UN mission, since riots in 2004, and commentators also noted the symbolism in the collapse of the ostentatious presidential palace as a result of the earthquake. However, in its emphasis on the complicity of the international community, the coverage also implied a moral obligation on the part of the rest of the world to help the beleaguered Caribbean nation and hence hailed a form of disaster citizenship premised on the need to intervene in the affairs of a failed state. For example, on January 15, as the scale of the disaster was becoming apparent, the *Daily Mirror's* front page featured a graphic image of an injured child under the headline "Help Us"—a very explicit injunction to care. In this context, much political analysis was devoted to the U.S. government response. Barack Obama's pledge of a $100 million aid package as well as thousands of troops to support the relief effort was seen as a much-needed contrast to George W. Bush's poor handling of Hurricane Katrina. U.S. media were keen to draw parallels between the two disasters. Chris Good, writing in the *Atlantic* politics blog, suggested that:

> The symmetry between Haiti's devastating earthquake and Hurricane Katrina is undeniable: the mass human suffering of an impoverished community, the shortage of vital resources, and the government response that was lacking—though for very different reasons—all point to the same understandings of civic responsibility that made Katrina a disaster for Bush and are motivating President Obama's forceful response to Haiti today. ("Haiti/Katrina symmetry: Biden goes to South Florida and New Orleans," January 14, 2010)

Other journalists saw the forceful U.S. response as a recognition of the country's close links to Haiti and the proximity and relevance of the country in geopolitical terms. *Guardian* reporters Ewen MacAskill and Richard Luscome wrote as follows:

> Haiti is close enough to the US to resonate for Americans in a way that disasters in Africa and Asia often do not. Haiti is only 90 minutes flying time from Miami. US television networks, which have abandoned much of their international coverage in recent years, sent in large crews and have been reporting almost non-stop. ("White House responds swiftly and generously after outpouring of public sympathy in America," January 14, 2010)

U.S. media coverage highlighted the role of the nation state with several stories stressing the role of its military in restoring order in Haiti. For instance, CNN reported that "Haiti's Port-au-Prince airport, now critical for the quick delivery of supplies and aid, was an uncontrolled 'mess' when the Air Force arrived Wednesday night to rehabilitate the facility, according to the commander of one of the Air Force's elite special operations units" ("Chaos at Haiti airport calmed by Air Force," January 15, 2010). Similarly, the Associated Press, among many other media organizations, reported on Obama's speech to the House Democrats, in which he made a case for U.S. exceptionalism and global responsibility:

> "This is a time when the world looks to us," Obama told House Democrats on Thursday. "And they say, given our capacity, given our unique capacity to project power around the world, that we have to project that not just for our own interests but for the interests of the world." ("Obama: won't tolerate excuses on U.S.' Haiti response," January 15, 2010)

Here, registers of disaster citizenship were inflected by geopolitical considerations, as the United States was celebrated for its "unique capacity to project power around the world." Obama called on forms of disaster citizenship which were cosmopolitan in their consideration of the "interests of the world" but remained narrowly national in the assured statement of the United States' continued world supremacy.

Thus, in both the Burmese and Haitian disasters, media coverage constructed a central role for the international community—and for particular nation states within it—which was justified in terms of assessments of the respective governments but in very different ways. The detailed engagement with the political contexts and histories in both cases demonstrates the complex ideological work done by media coverage in the elicitation of disaster citizenship. Here, Western liberal democratic notions of the state belong to the sphere of consensus, whereas arguments around interventions in the operations of sovereign nation states belong to the sphere of legitimate controversy and hence require extensive justification and elaboration.

By contrast, in the case of the Sichuan earthquake, the attitude of the international community, reinforced by China's perceived strong handling of the disaster, was that there was little need for assistance from the international community:

> We're here if you need us.
>
> Despite international focus on the devastation in Myanmar, that's the message international aid organizations have given to China, a country thought to have the domestic resources necessary to respond to the earthquake.
>
> "They're well organized, and they've developed good response capacities because they do have a history of earthquakes and other disasters in China," said Stephanie Bunker, spokeswoman for the United Nations Office for the Co-ordination of Humanitarian Affairs. ("Aid groups on alert to help Beijing—if asked," *Globe and Mail,* May 13, 2008)

In this story, the emphasis on the "domestic resources" once again foregrounds the context of the nation state and, as a corollary, the idea that if a national government is willing and able to meet its assumed responsibilities, there is no need for "outside" assistance. This response stands in stark contrast to the ways in which the Burma cyclone was discussed. Indeed, the China earthquake and the Burma cyclone occurred almost simultaneously and therefore gave rise to much coverage that compared the two countries, reflecting on differences in response, resources and communication, among other things. A Nexis UK search found that in the two weeks after the earthquake hit Sichuan, there were 132 stories in major international newspapers discussing the two disasters together (using the search terms "Burma" and "cyclone" and "China" and "earthquake" for the period from May 10 to May 24, 2008). These stories shared a metanarrative that was, first of all, premised on the universal values of democracy and human rights, and accordingly, condemned the undemocratic nature of the respective regimes, and second, assessed the actions of the regimes under the assumption of state responsibility. A BBC News online report, written by Paul Danahar, compared the official response in Burma and China and was representative of prevailing discourses. Danahar suggested that, "The generals in Burma find themselves accused of an 'inhuman' response to their disaster, bordering on a 'crime against humanity.' In contrast China has found itself in the unusual position of being showered with international praise, both for its reaction to the disaster and the openness with which it has allowed the details to be reported" ("Burma and China: Tale of two disasters," May 19, 2008). Danahar's account aptly captures some of the key metanarratives: The assumption of state responsibility appeared so ingrained that in rejecting it, Burma's military government was guilty of "inhuman" actions or crimes against humanity. China, by contrast, surprised the world not just in responding to its responsibility but also in its openness—a key democratic principle, which is otherwise rarely used to characterize the leadership of the world's most populous nation. An opinion piece published in the Australian *Advertiser* newspaper, titled "Contrasting tale of two disasters," drew on similar themes in comparing the responses of the two "totalitarian, non-democratic" states, which share "a sorry track record of ruthless repression and placing ideological considerations above the welfare of their citizens" (May 16, 2008). Here, as in other comparisons between the two disasters, what is foregrounded is, on the one hand, the ways in which the two regimes both run counter to the Western consensus around the value of liberal democracy (and, by extension, universalized values around human rights), and on the other hand, the marked differences in how the regimes responded to the disasters:

> Both regimes have an unfortunate history of showing arrogant disdain or indifference when members of the international community have attempted to call them to account for a multitude of human rights abuses or uncaring practices towards their long-suffering peoples.
>
> However, the latest natural disasters have produced very different reactions from the authorities in these two tightly controlled regimes.
>
> On the one hand, the lethargic, incompetent and callous reaction of Burma's bumbling generals has sparked a groundswell of international outrage.
>
> On the other, the swift and well-organised response of China's leaders to the crisis has earned that nation well-deserved acknowledgement and even admiration. . . .
>
> With the Olympics less than three months away, the Chinese authorities have through their elaborate and professional relief efforts been anxious to convince the world of two things: First, that they are in control and, second, that China is a vibrant and modern country on the way to opening up to the rest of the world.
>
> Burma's military junta leader, Senior General Than Shwe, appears hell-bent on sending exactly the opposite message to the world, causing untold suffering for his compatriots. (*Advertiser*, "Contrasting tale of two disasters," May 16, 2008)

The report's mention of the upcoming Olympic Games in Beijing exemplified how both longstanding and short-term political contexts were drawn into the disaster coverage and reflected broader debates about China's emergence as a player on the global political stage. It also hailed particular audience reactions and, hence, forms of disaster citizenship, based on constructions of the international community's reactions: While it invited the reader to share in the well-deserved acknowledgment and even admiration for China's success in addressing the disaster, it equally encouraged the reader to become part of a groundswell of international outrage. This story, then, set out the parameters of acceptable reaction, hailing the reader as part of an *international* community to pass judgment on the respective regimes on the basis of an awareness of the relevant geopolitical contexts.

Indeed, coverage of the Chinese earthquake also highlighted the geopolitical and economic importance of China by drawing attention to the global financial impact of the earthquake. For example, a story from the *International Herald Tribune* was one of many to focus on the earthquake's reverberations in the financial markets:

> Chinese stocks gave up some of their gains after a strong earthquake in southwest China rocked buildings in Shanghai just before the market closed. The yuan slipped against the U.S. dollar.
>
> After the earthquake, several high-rise buildings in Shanghai's financial district were evacuated, disrupting trading.
>
> The exchange in Shanghai said the market was functioning normally until its regular close.
>
> "The quake made investors panic," said Qiang Xiangjing, an analyst at Citic-Kington Securities. "Most of them have not yet figured out what has happened, including what losses the quake may cause." ("Falling yen and bank takeover talks help shares," May 13, 2008)

The emphasis on market consequences, which was also strongly in evidence after the 2011 Japanese earthquake, demonstrates that in a globalized environment, disasters are not just understood in terms of their impact on the immediate and local surroundings but also in terms of how they affect an increasingly interconnected global financial system. This could be seen as another form of domestication of disaster news, insofar as such stories made the disaster relevant—albeit in a way that distanced audiences from the suffering of victims—to audiences around the world. However, this type of story also constructed the audience as consumers situated within a liberal and globalized economy, rather than as compassionate citizens of the world. Nonetheless, it fed into a larger narrative around the transformation of China from a backward and repressive dictatorship into a dynamic and open world power. The openness of the country—particularly in comparison to the Burmese example—became a major theme of coverage, as in this example from the *Daily Telegraph:*

> MINUTES after the earthquake struck in Sichuan yesterday, the internet was alive with videos and witness accounts of the disaster.
>
> It was a contrast to the events in Tangshan 32 years ago, when the Chinese government

> refused for months to admit the 7.8 magnitude earthquake had even happened, despite the deaths of an estimated 255,000 people.
>
> China's new-found transparency has much to do with the arrival of the Olympics in Beijing in August. Although there are at least 29 journalists in Chinese prisons, 19 of whom are bloggers, the country has had to relax its reporting restrictions as a condition of holding the Games. ("Unlike 1976, news was out instantly," May 13, 2008)

While this story framed the new openness as a result of the upcoming Olympic Games, another story, in Australia's *The Age,* suggested that the realities of technological change have forced party leaders to face bad news head-on. The article quoted a former reporter for the official Xinhua news agency, who suggested that "nowadays—you can't hide anything. Nothing stays a secret with the internet" ("You can't hide a major disaster in today's China," May 14, 2008). This story, while attributing the change to external circumstances, signaled a profound transformation of Chinese politics in the direction of Western liberal ideals of free speech. This poignant theme in coverage, as we explore in the next section, gave rise to the idea that the openness and decisiveness of the state created an unprecedented atmosphere of national pride and civic responsibility, even as critical voices also came to the fore. Like never before, Chinese citizens were allowed into the global public sphere.

Citizens, Free Speech and Democracy

Stories about disasters do not merely convey understandings of *audiences* as citizens but also, as we discuss in more detail later, about the *disaster victims* as citizens. Victims of the disaster were politicized as citizens when they were given voice in stories about the disaster to illustrate the limitations or merits of the aid effort. In all three cases, citizens were constructed in ways which backed up the dominant narrative around the role of the nation state in responding to the disaster: In China, citizens reflected two divergent political assessments in the aftermath of the disaster. On the one hand, the coverage reflected a new mood of national pride and togetherness, while, on the other hand, it gave voice to criticism of poor construction standards, which led to the death of thousands of children in school collapses. In Burma, the brave citizens willing to give interviews to foreign journalists were few and far between, but they readily backed up the prevailing narrative around the inhumane response of the military dictatorship. Finally, in Haiti, coverage allowed only citizens of the beleaguered nation state to voice criticism of its government's handling of the situation, while outside observers—as we have already indicated—were confined to observing long-standing, structural reasons for the regime's failure to competently handle the consequences of the disaster.

In the Sichuan earthquake, amid the widespread celebration of the government's decisive action, the angry voices of bereaved parents began to emerge as it became clear that shoddy construction standards were to blame for the many school collapses. The anger of parents was covered widely—it demonstrated, as we further explore in the next chapter, that those affected by disaster are granted a significant political expression entitlement and, more concretely, that anger is a powerful political emotion, which serves a key role in raising issues of blame and responsibility.

Indeed, the feelings of bereaved parents were used to highlight what was generally a far more prominent discourse in coverage of Burma and Haiti, that of the blurring between "man-made" and "natural" disasters. A front-page story in *The New York Times* made this point explicitly by drawing on the voice of a bereaved parent: "This is not a natural disaster," said Ren Yongchang, whose 9-year-old son died inside the destroyed school. His hands were covered in plaster dust as he stood beside the rubble, shouting and weeping as he grabbed the exposed steel rebar of a broken concrete column. "This is not good steel. It doesn't meet standards. They stole our children" ("Chinese are left to ask why schools crumbled," May 25, 2008). An article in the *Sunday Telegraph* reported from the small town of Wufu, where nearly all of the casualties of the earthquake were children between the ages of 10 and 13 who perished when the Fuxin 2 Junior School collapsed:

> But if grief is the dominant emotion in Wufu, then anger runs it a close second. As you walk through this rural community, it is impossible not to be struck by the profusion of home-made banners strung across streets and hanging from walls. They point to a story of corruption and incompetence that the bereaved believe did more to kill their children than any natural disaster.
>
> One reads: "Our children are not dead directly due to the earthquake, but because of a tofu building." ("Grief and anger in town of dead children," May 15, 2008)

The voices of bereaved parents were thus heard through a variety of channels—calling officials to account, speaking to foreign journalists and expressing their anger on homemade banners—and underpinned a critical discourse amid a sea of praise for the Chinese government.

That critical voices were heard in global media was in itself remarkable given the Chinese government's history of suppressing freedom of speech. Alongside the emergence of critical voices, the perceived new openness of the regime, and the ways in which it gave rise to new and more democratic notions of citizenship, were key themes not only in the general disaster coverage but also in stories about Chinese citizens and their reactions to the disaster. In this, acts of citizenship and belonging became a synecdoche for sweeping cultural and political transformations. Several reports thus reflected on the unprecedented civic-mindedness of Chinese citizens and its relationship to the equally unprecedented extent of media coverage of disaster, as in this example from Singapore's *Straits Times:*

> With Beijing ever wary of any sort of activism not initiated by the state, this nationwide, ground-up outpouring of grief and action is all the more remarkable, note observers. The local press is replete with touching stories, like that of a dishevelled beggar who showed up at a donation box repeatedly, putting in his meagre takings each time. A week after the quake, the *Shanghai Daily* newspaper reported that more than 400 residents of the city of glam and glitz had registered to adopt children orphaned by the quake. White-collar types are selling their treasured art pieces or much sought-after Beijing Olympics tickets to raise money for quake victims. . . .
>
> A huge grassroots response is perhaps natural, given that this disaster is the worst in decades and its "blood-soaked reality" was beamed straight into people's homes and consciousness, noted Professor Hu Xingdou, of the Beijing Institute of Technology.
>
> Indeed, the media—and Beijing's relative openness in news coverage this time—played a key role in mobilisation. . . .
>
> And now that regular Chinese have had the epiphany that they have a role in society—beyond the traditional one of caring for their own families—they might want more space for activism. ("Earthquake awakens civil activism in China," May 23, 2008)

By contrast, reporting from Burma stressed the scarcity of reliable information available and the difficulty of accessing citizens' perspectives. Many of the stories from Burma relied on information from the state-run television, international news agencies and expat Burmese, while relatively few were from journalists actually inside the country. Burma emerged as an insular state of isolated and fearful victims who had never enjoyed the experience of being addressed as citizens and were largely bereft of agency. One story from the British tabloid, the *Sun,* dramatized the fear and isolation of Burmese villagers:

> One villager whispered: "If you disobey them they will tear out your fingernails one by one. Do as we do—do as they say."
>
> A local said: "The dead people swept into the water have gone from around here now. The soldiers never came to move them. We did it ourselves. Everyone is hungry and scared of disease but no one is coming to help." ("Burma's starving, but cops are busy…nicking Sun men," May 15, 2008)

Another story, published in the British *Daily Telegraph,* revealed rare "glimpses of dissent" against the "cruel, power hungry and dangerously irrational" military government ("Out of this tragedy some light may begin to shine on Burma," May 7, 2008). In doing so, it drew on the voices of the cyclone's victims to underpin assessments of the Burmese government. It implied that any open criticism of the junta was courageous, dangerous and extraordinary:

> On Sunday, two soldiers demanding a gallon of petrol, which is in short supply, from a Rangoon housewife. "I'll give it to you because you are people asking for my help," she told them, "but I hate this uniform!" A group of neighbours stood about in silent support. "I felt like clapping," said one witness. ("Out of this tragedy some light may begin to shine on Burma," May 7, 2008)

Overall, however, Burmese citizens' voices were few and far between, reflecting the difficulties of access. Most critical perspectives came instead from overseas Burmese, as in a *Toronto Sun* story that cited a Canadian Burmese woman who was concerned about the fate of her relatives, stating the following: "I'm scared for them. The military government don't take care of the people" ("Local Burmese fear for kin in wake of storm," May 6, 2008). Though limited in numbers, the Burmese voices added to the chorus of condemnation directed at the military regime for absconding from its responsibility to assist its own citizens.

In the case of Haiti, citizens seemed authorized to say what other foreign nationals could not, even if their statements backed up the discursive consensus on the failed nation state: Because of the global power dynamics shaping the discourses surrounding this disaster, it appeared to be permissible only for Haitian citizens to criticize the absence of their own government in the aid effort. A *Los Angeles Times* story illustrated the country's poverty in poignant detail and described the anger of residents about the lack of government assistance:

> They were furious, though not surprised, that they were left to themselves to dig out the trapped, haul off the dead, beg for help for the dying.
>
> Hubert Benjamin, 59, blamed his own government and figured that it would squander any international aid it received.
>
> "I know if they give it to them to give to the Haitians. I know already they won't give it to us."
>
> "Look at how many people die here on the ground. No one comes to see them. Right now there is still someone crying in a building down there." ("Lack of aid leads to anger among Haitians," January 15, 2010)

The journalistic practice of allowing victims the opportunity to critique government for its inaction is not, as we discuss in more detail in the next chapter, unique to this disaster. However, what was distinctive about this case was the fact that external observers were only "allowed" to consider structural issues. A weak or nonexistent state, it would seem, is not viewed as a fair target of external criticism. What united all three cases was the use of citizens' voices to generate explicitly politicized empathy for those affected by the disaster, whether in stories about Chinese towns and cities robbed of their children, the hunger of neglected Burmese cyclone victims or Haitians' despair at their government's inability to assist in the relief efforts.

Conclusion

Despite the globalized and transnational nature of media and the disasters they cover and their potential to elicit cosmopolitan sensibilities through journalistic witnessing, the nation state and its place within the geopolitical landscape of the world remain highly salient. The coverage we examined reinforced dominant ideological formations and discourses but also fed into ongoing political debates about issues including the injustices of the military dictatorship in Burma, the changing place of China in the world and the complicity of the world's wealthy countries in the poverty and devastation of Haiti. Disaster coverage, then, is a lens through which we can analyze the always-dynamic global political landscape and the place of citizens within it.

In all three disasters, it appears that the underlying story was about the role of the state and the international community in responding to these devastating disasters and aiding the beleaguered victims. Citizens of Western democracies were, through the metadiscourses of democracy embedded within the sphere of consensus, reminded of their own privilege but also of the plight of victimized citizens whose lives, already blighted by failures of governance, were touched by disaster. Through this process, the political situations in Burma, Haiti and China were brought, at least temporarily, to the forefront of the news agenda.

Thus, the chapter has demonstrated that while disaster coverage can build cosmopolitan empathy, the cosmopolitan imagination is constructed through and bounded by the nationally and geopolitically inflected narratives of disasters. We have further refined these arguments by suggesting that the discourses around forms of cosmopolitan citizenship are not fixed but instead constructed and justified in the context of specific disasters. In the case of the forms of disaster citizenship studied in this chapter—which are themselves radically historically specific—citizens of nation states and the international community were called to action to help the citizens of another nation state. Further—and along the lines of the points made earlier about disaster citizenship as a particularly reactive and negative form of citizenship—citizens appear to be hailed more strongly to address their attention to particularly intractable problems, problems that have proven impossible to solve within the remit and boundaries of the nation state. Disaster citizenship is *inter*national rather than *trans*national because it continues to depend on the framework of assumption of state responsibility. It is only when the nation state fails in this responsibility that the international community, as represented by national and international actors staking out their claims in part through media coverage, steps in to call for interventionist measures. As was evident in the coverage of the disasters in Haiti and Burma, the legitimacy of intervention is not taken for granted, insofar as it is understood as an incursion on national sovereignty, which therefore places it within the sphere of legitimate controversy. Intervention appears to require lengthy and substantial justification on the basis of such absolutist concepts as brutal dictatorships and failed states.

What emerges here, then, is a complex picture of the possibilities and limits of cosmopolitan citizenship: While there is a clear scope for the cultivation of cosmopolitan sensibilities in some forms of disaster coverage, it is also the case that the registers of disaster citizenship, as identified here in Western media coverage of Burma, China and Haiti, construct audiences as citizens who may well experience solidarity and empathy in relation to distant sufferers. But their primary mode of acting on these emotions is to back up the actions of their nation state and humanitarian aid organizations to come to the rescue of those afflicted by the disaster and in need of help from the world beyond the boundaries of their own nation (see also Chouliaraki, 2006). Disaster citizenship, then, does not necessarily represent the globalization of rights and empowerment envisioned by proponents of cosmopolitan citizenship, but it does construct spectators who are able to care for distant and different others whom they will never meet but whose plight is too terrible to ignore.

8

Anger and Accountability in Disaster News

IN THIS CHAPTER, WE EXAMINE HOW MEDIATED DISASTERS CAN GENERATE opportunities for political challenge and opposition. In particular, we analyze the role of anger in constituting citizenship and moving people into new formations and forms of political action—or inaction. As discussed in Chapter 3, we assume that the media play a central role in facilitating citizen participation and deliberation. Going beyond a view of anger as a "negative" emotion leading to acts of violence, or merely a psychological phenomenon, we examine it as performative political speech, which produces relationships with others and may move people to express dissent and make claims for justice. To become a politicized emotion, anger needs to be collective and named and recognized in public (Lyman, 2004). As Simon Thompson (2006, pp. 128–129) wrote, anger needs a "framework of interpretation" in order to be transformed into political action and lead to collective struggle against injustice.

In this chapter, we see disaster reporting as potentially providing such an interpretative framework. Whereas the previous chapter considered the ways in which disaster coverage embeds politicized narratives around the power and responsibilities of states and other institutional collectivities, this chapter looks at how disasters politically matter at the level of citizen action; it reflects on how they can give voice to angry victims, and, in the process, smuggle in politically charged questions of blame and accountability. As we discussed earlier, emotion discourses in media narratives are crucial for negotiating the meaning of disasters as well as making moral judgments that involve deciding on a course for political action. In more general terms, mediated collective emotions may include significant potential for social change at a time when traditional politics has become less relevant to many. Emotion may be a powerful source of political action and social change, but it can also be a source of inaction and stability (Gould, 2009).

In Chapter 6, we investigated the manifestations and political repercussions of compassion in the public sphere. Here, our focus is on the role of anger and blame in disaster coverage. Anger is intrinsic to the processes of attributing responsibility and blame, which are central to disaster coverage. We argue that the discourse of anger in news media is crucially connected to the capacity of mediated disasters to contribute to political contestation and challenge. Anger and moral outrage on behalf of disaster victims can be a powerful motivation for dissent when there is someone—an individual, group or institution—to blame for the injustice. Theorists of social movements have suggested that protests emerge from a "moral shock," which motivates people to react to a "perceived injustice" (Goodwin et al., 2001, pp. 16–17; Jasper, 1997, p. 106). The media provide a forum for articulating the anger of the afflicted, thus representing the voices of groups and individuals who are otherwise marginalized within the media's discursive universe, but through emotive coverage and a "blame frame," the media can also sustain and deepen public anger to the extent of forcing a change in political response. In his book, *Civil Society and Media in Global Crises,* Martin Shaw (1996) discussed how British television news coverage of the Kurdish refugee crisis in 1991 had a "remarkable effect on Western governments' policy. By weaving together a powerful moral and emotional package, sizzling with anger (that of the victims and journalists as eyewitnesses), news programs put the victims' accusations against the powerful and clearly attributed the responsibility to Western governments. As Shaw argued, the "graphic portrayal of human tragedy and the victims' belief in Western leaders was skillfully juxtaposed with the responsibility and the diplomatic evasions of those same leaders to create a political challenge which it became impossible for them to ignore" (p. 88).

Both compassion and anger are "moral emotions" in that they highlight moral aspirations and norms about "what should be" (Aminzade & McAdam, 2001, p. 19). Compassion and anger, however, open up a possibility for different forms of political communication and action to emerge in the public sphere. While compassion orients us to relieving the suffering of others, anger is directed at the source of a tragedy, typically at those authorities perceived to be responsible. Public compassion and mourning are emotionally powerful and lasting narratives in disaster news and can be understood in terms of community building, since expressions of grief, mourning and sympathy are easily appropriated by the media to construct a national consensus of feeling and, consequently, contain anxiety and dissent (e.g., Pantti & Wieten, 2005; Thomas, 2002). Moreover, the discourse of grief, appropriated particularly forcefully by tabloid media, is easily recognized as addressed to "us," that is, all "we" who have suffered a loss can recognize ourselves as the ones addressed (cf. Warner, 2002, pp. 76–97). Anger, on the other hand, has been seen to have a particular political significance because it is integrally related to issues of power, justice and collective activism and can contribute powerfully to mobilization (e.g., Goodwin et al., 2001; Holmes, 2004; Jasper, 1997; Lyman, 2004; Muldoon, 2008; Thompson, 2006). The political value of anger, then, has been seen to lie in its judgmental nature, in its capacity to communicate that an injustice has been committed and, consequently, its ability to challenge existing power relations (e.g., Lyman, 2004). As Elizabeth Spelman (1989) wrote, the expression of anger is an act of defiance through which the people claim the right to become a judge of the powerful.

It is precisely because of this potentially defiant nature of anger that the elite has traditionally tried to suppress angry reactions, especially when it comes to the anger of subordinate or marginalized groups (Lyman, 2004; Spelman, 1989; Stearns & Stearns, 1986). However, as Peter Lyman (2004) argued, it is essential for a constructive politics that angry speech is heard rather than silenced or domesticated: "The benefit of taking anger seriously is that listening to those who feel they have lost their right to be heard reduces social suffering, enriches political dialogue, and enhances the ability of politics to redress injustice" (p. 133). It has been claimed that disasters tend to be reported according to narrative and ideological conventions which serve to naturalize their occurrence by presenting them beyond the control of people and hiding the fundamental structure of injustice behind them (e.g., Hammock & Charny, 1996; Tester, 2001).

In this chapter, we are interested in how expressions of anger in media coverage may influence the "blame game" and, moreover, may serve as a means of "denaturalizing" disasters. Shanto Iyengar (1991) proposed that "episodic" news reports, in which problems are depicted by focusing on the circumstances and experiences of specific individuals, foster individualistic attributions of responsibility, while "thematic" news stories, emphasizing social conditions with general facts and figures, would lead the audience to ascribe more blame and responsibility to structural causes. His work belongs to the line of research that studies how the way news events are framed and narrated influences the audience's attribution of responsibility for the social consequences of those events. In the context of disaster reporting, then, employing a thematic frame would lead the audience to assess the disaster in terms of external structural factors, such as the actions of the government, while reporting on the consequences of the disaster through the eyes of victims would decrease the sense of government responsibility and lead the audience to ascribe more responsibility to the affected persons (Ben-Porath & Shaker, 2010). Studies in this field have confirmed that subtle cues in news messages, such as images and expressions in news presentations (e.g., Knobloch-Westerwick & Taylor, 2008), influence news consumers' attributions of blame. Studying the effects of race in news images on the attribution of responsibility for the outcomes of Hurricane Katrina, Ben-Porath and Shaker (2010) concluded that the inclusion of images of Katrina's victims tended to reduce white readers' sense of government responsibility but did not lessen black readers' perception that the consequences of this humanitarian crisis were "a product of government incompetence or indifference in the face of suffering of an overwhelmingly Black population" (pp. 482–483).

Some mediated national and global disasters may open up a space for expressions of anger and, consequently, provide symbolic and discursive resources for political challenge and dissent. Indeed, some disasters are powerfully framed in terms of anger and critique, while in some such frames are largely absent (Pantti & Wahl-Jorgensen, 2011). When expressions of anger are present in disaster coverage, they may come in a variety of forms. First, there are direct expressions of anger from news sources: Journalists report on emotions through quotes, allowing the news sources to express their anger directly, thus remaining "objective." Second, some disaster coverage includes indirect descriptions of anger, which occur when journalists report on emotions by interpreting individual and collective emotions (for example, in making references to the public mood). Finally, some disaster coverage draws on "authorial emotions," which come into play when journalists' own emotions are expressed.

The Blame Game in Disaster Reporting

Sociologist Charles Tilly (2010) suggested that giving credit and blame uses the universal tendency to describe social experiences as stories or simplified cause–effect accounts. Blame-giving follows a logic that moves from the magnitude of the change or negative outcome to identifying an agent who helped bring it about and judging the agent's competence and responsibility for the action that produced the outcome. In their large-scale content analysis of press and television news stories, Semetko and Valkenburg (2000) found that the "attribution of responsibility" frame, along with the "conflict frame," were the most frequent ones in the news. Moreover, they demonstrated that quality news outlets were more likely to adopt the responsibility attribution frame than tabloid news media. Disaster reporting necessarily presents or implies responsibility attributions and sometimes thereby opens up for political debates on accountability.

When a disaster occurs, finding a "suitable" culprit might be a condition for a consensual media ritual where media act in concert with the government. In her discussion of "disaster marathons," Tamar Liebes (1998) wrote the following:

> Stories of disaster invite a hermeneutic search for the culprit, someone to whom to assign the blame. The less possible it is to point to the actual villain, the less the chance of satisfactory resolution, and the more powerful the role of television in providing the framing. Thus when the leader's assassin is caught and the murder is declared as the act of one "mad" individual, journalism can relax into the "priestly" mode. When two Israeli army helicopters crash, killing seventy-two soldiers, the question of responsibility may remain unsolved, and the live broadcast may be responsible for creating a climate in which public spirit will turn into rage (against the air force, the army or policy makers sending soldiers to Lebanon) or tamely go into collective mourning. (p. 74)

Disasters can support the political status quo but also accentuate political crises, and this is particularly true when the blame and responsibility for a disaster can be given to those in power. Some disasters evidently do not conform to the rituals of national integration and solidarity but may shift public attitudes and upset the political status quo by attracting negative attention to current policy (Birkland, 1997; Cottle, 2009a). As we have argued, not only "man-made" disasters but also "natural" ones can been seen as a result of human activity or negligence and therefore as an instance of (nondivine) injustice. One can be considered responsible and blameworthy without having "caused" an event. Along those lines, as we saw in Chapter 6, governments and officials can be held responsible for not doing enough to prevent the disaster from happening or limiting the damage it causes, or they may be blamed when their disaster response is deemed slow and ineffective. In times of crisis, political leaders are expected to lessen the impact of the crisis (Boin, 't Hart, Stern, & Sundelius, 2005). Even in a situation when public expectations of politicians' performance are generally low, a crisis generates high expectations. This is because, as Lars Nord and Jesper Strömbäck (2009) have argued, in a crisis situation "options for acting appropriately are generally considered more limited" (p. 37). Mediated disasters, then, can reinforce confidence in government but they may also accentuate political crises when the response is seen as inadequate. Perceived shortcomings in disaster response drive governments into a defensive position, which makes them even more vulnerable to criticism and negative public evaluation.

In major crises and disasters, however, the norm has been for the news media to show unity with the government and work to strengthen shared values by temporarily abandoning the objectivity norm in favor of addressing the suffering and calling on the national "we" (Schudson, 2002, pp. 40–41). In some circumstances, this normal sphere of consensus in the aftermath of major national disasters is replaced with the "sphere of controversy" (Hallin, 1986), where the media's ritual consensus with political institutions is absent or "decentered" (Durham, 2008). Crisis events may also represent moments in which journalists achieve power in relation to the political establishment. Because of the increased use of Internet and mobile communications for receiving and sharing information, today's journalists are less dependent today on establishment sources.

News reporting of the Asian tsunami and Hurricane Katrina illustrates how disasters can open up opportunities for political challenge and opposition. At the same time, coverage of the 2008 Sichuan earthquake demonstrates that in cases where the government's handling is broadly constructed in positive terms, those affected by the disaster are seen as entitled to raise issues of blame and accountability. Much of the Asian tsunami coverage in Nordic countries focused on the blame and public anger in the face of what was perceived as governmental authorities' inept handling of the disaster. In Sweden, the "blame game" coverage reinforced the low public confidence in politicians by suggesting that the crisis management had failed to meet the expectations of the people (Nord & Strömbäck, 2009). Criticism of the handling of the disaster in Sweden by politicians and officials became a prolonged national political crisis, leading to the defeat of Prime Minister Göran Persson's Social Democratic Party in the general election in 2006. As Ullamaija Kivikuru and Lars Nord (2009) wrote, "The initial combination of the absence of effective governmental information routines, delays in press conferences and news media comments and perceived lack of compassion and strength among the political leadership, all proved to be devastating when the public evaluated Swedish governmental performance during the tsunami disaster" (p. 12). The authors concluded that the stable Swedish political system, based on power-sharing structures and sophisticated arrangements for determining liability, "may prove insufficient when faced with deviations from normality and routines" (p. 13).

In Finland, the government's crisis management was also strongly criticized in the media, but the "blame game" did not lead to a similarly devastating public evaluation of the government's performance. Instead, the consensus was rapidly restored and the tsunami response was not among the themes discussed during the next election campaign (Jääsaari, 2009; Kivikuru & Nord, 2009). In her analysis of tsunami communications, Johanna Jääsaari (2009) employed the concepts of "media events"

(Dayan & Katz, 1992) and "disaster marathons" (Liebes, 1998) to explain how the consensus could be so rapidly restored in Finland despite the media criticism and public anger in the face of the breakdown of official communication. During the disaster marathon phase, when the crisis was still acute, the media took every opportunity to criticize the Finnish government's rescue efforts and mismanagement of communication. Finnish media monitored Swedish media and used similar examples of shortcomings of the authorities to evaluate the Finnish authorities' performance. The "de-centered media ritual" (see Couldry, 2003, Durham, 2008) was recentered through a set of media events arranged by the government and sponsored by national television (many broadcast live) with the prime minister and the foreign minister as the leading actors. In press conferences, the government succeeded in convincing its citizens that the situation was under control and reinstated the relationship between the media and official sources. In addition to press conferences, various ceremonies marked the event as one of national sorrow and mourning. The live coverage and public mourning affirmed the bond among the government, the media and the public and, most importantly, returned the meaning-making power (or the power to define the government's response) to the official authorities. Through these media events, the government, the media and the people were united in a consensus that "the crisis was being solved in a way that honored the traditional 'Finnish' values of efficiency, equality and fairness" (p. 61). Drawing on the written texts written by members of the tsunami audience panel, Jääsaari concluded that "for the TV-watching public, this was far preferable to the unsettling and distracting atmosphere created by the media's criticism of the crisis management" (p. 77).

In the case of Hurricane Katrina, both the mainstream media in the United States and the international media also gave vent to the expressions of anger over time. News media around the world opened up a space for the critical elaboration and framing of this disaster. The president of the United States, George W. Bush, was identified by some as a principal source of blame, not heeding advance warnings and then unthinkingly commending state officials "for doing a great job." By such means, Hurricane Katrina exposed the normally invisible inequalities of "race" and poverty in American society and became an opportunity for political appropriation by different projects and discourses worldwide. The BBC online news website, for example, positioned itself as a portal for world opinion, exhibiting opinion pieces from newspapers from around the world and providing hyperlinks to many of them in a piece titled "World press: Katrina 'testing US'" (September 5, 2005, http://news.bbc.co.uk/1/hi/world/americas/4216142.stm). It is instructive to consult just a few of them here:

> Bush is completely out of his depth in this disaster. Katrina has revealed America's weaknesses: its racial divisions, the poverty of those left behind by its society, and especially its president's lack of leadership. (Phillipe Grangereau in France's *Liberation*)
>
> Hurricane Katrina has proved that America cannot solve its internal problems and is incapable of facing these kinds of natural disasters, so it cannot bring peace and democracy to other parts of the world. Americans now understand that their rulers are only seeking to fulfill their own hegemonic goals. (Editorial in Iran's *Siyasat-e Ruz*)
>
> Co-operation to reduce greenhouse gas emissions can no longer be delayed, but there are still countries—including the US—which still do not take the issue seriously. However, faced with global disasters, all countries are in the same boat. The US hurricane disaster is a "modern revelation," and all countries of the world including the US should be aware of this. (Xing Shu Li in Malaysia's *Sun Chew Jit Poh*)
>
> Katrina is testing the US. Katrina is also creating an opportunity for world unity. Cuba and North Korea's offer of sympathy and aid to the US could also result in some profound thinking in the US, and the author hopes that it will not miss the opportunity. (Shen Dingli in China's *Dongfang Zaobao;* http://news.bbc.co.uk/1/hi/world/americas/4216142.stm)

Differences of geopolitical interests and cultural outlooks clearly register in these very different views from around the world and here relayed via BBC online news onto the global news stage. The exposure of America's continuing racial divides and depth of poverty by the hurricane appears to have sullied its projected international image as a "free democracy" for some observers. Countries normally regarded as political pariahs or as economic supplicants by the U.S. government turned the tables and offered their support to the world's mightiest power in its evident failure to respond to its home-grown humanitarian disaster. And yet others took the opportunity to make the connection to climate change and the irony of the U.S. position of refusing to sign the Kyoto treaty. Indeed, such was the mounting criticism played out in the news media that commentators began to speak of George W. Bush's "Katrinagate." In such ways, then, Hurricane Katrina was staged as a global focusing event, in which global voices of criticism and dissent, as well as sympathy and support, entered the media frame.

Criticisms of city officials, failed evacuation plans, inadequate relief efforts and identification of bodies, the apparent abandonment of some of the poorest people in American society to their fate as the militarized response to the aftermath, were all voiced in the national news media as well (Bennett et al., 2007; Durham, 2008; Robinson, 2009; Rojecki, 2009). As Kitch and Hume (2008, p. 37) wrote, "Families of the deceased moved from a state of confusion and grief to anger, and the press reported that progression." Frank Durham (2008) examined the television coverage of Hurricane Katrina as a populist or "decentered" form of media ritual. Rather than acting as the upholders of dominant social norms, television journalists, in the absence of federal government sources, appealed directly to the people in an effort to focus the government's attention on the effects of the disaster on the people of New Orleans. Durham drew on the example of CNN reporter Anderson Cooper's angry outburst on national television when interviewing a Louisiana senator about the government's response to the disaster to illustrate the dominant frame or meta-narrative of anger at the government. The reporter identified with the anger and pain of the afflicted, attacking the official source for abandoning the citizens and placing himself in opposition to his source "in a way that resonated culturally with viewers" (p. 109):

> Cooper: Joining me from Baton Rouge is Louisiana Senator Mary Landrieu. Senator, appreciate you joining us tonight. Does the federal government bear responsibility for what is happening now? Should they apologize for what is happening now?
>
> Landrieu: Anderson, there will be plenty of time to discuss all of those issues, about why, and how, and what, and if....Let me just say a few things. Thank President Clinton and former President Bush for their strong statements of support and comfort today. I thank all the leaders that are coming to Louisiana, and Mississippi, and Alabama to our help and rescue. We are grateful for the military assets that are being brought to bear. I want to thank Senator Frist and Senator Reid for their extraordinary efforts. Anderson, tonight, I don't know if you've heard—maybe you all have announced it—but Congress is going to an unprecedented session to pass a $10 billion supplemental bill tonight to keep FEMA and the Red Cross up and operating.
>
> Cooper: Excuse me, Senator, I'm sorry for interrupting. I haven't heard that, because, for the last four days, I've been seeing dead bodies in the streets here in Mississippi. And to listen to politicians thanking each other and complimenting each other, you know, I got to tell you, there are a lot of people here who are very upset, and very angry, and very frustrated. And when they hear politicians slap—you know, thanking one another, it just, you know, it kind of cuts them the wrong way right now, because literally there was a body on the streets of this town yesterday being eaten by rats because this woman had been laying in the street for 48 hours. And there's not enough facilities to take her up. Do you get the anger that is out here?

This exchange was widely reproduced and discussed, precisely because it was relatively unusual in the explicit expression of the journalist's anger, in his role as the representative and advocate of the people. But even in disasters where government handling is widely praised, there is still room for questions of blame and accountability. In the 2008 Sichuan earthquake, discussed in more detail in the previous chapter, media coverage drew on the angry voices of parents in blaming shoddy construction practices for hundreds of school collapses, which emptied villages and towns of its children. Amid celebrations of the swift and decisive actions of the Chinese government, and the emerging—and novel—civic-mindedness of Chinese citizens, the anger of parents became a major theme in coverage. This article from the UK *Times* newspaper thus highlighted the role of parents in calling government to account for the poor construction standards:

> In almost every town and village rocked by China's massive earthquake distraught parents point to the ruins of a school where their only child died. Angry citizens have deluged the Government with demands for an explanation and the Government vowed yesterday to punish anyone found to be responsible for shoddy construction....In a rare, real-time online exchange with ordinary Chinese, officials' measured answers were met with the kind of furious comments that have echoed across the internet since the quake left whole villages destroyed. One said: "China's Government buildings at every level are more magnificent than those of developed countries, the schoolrooms are worse than Africa's, who's to blame!!!" Education officials in many provinces made promises to tear down and rebuild schools if they were not earthquake-safe. "There may have been shoddy work and inferior materials during the construction of some school buildings," one official said. (May 17, 2008, p. 42)

In the Sichuan earthquake, the voices of bereaved and angry parents were heard through a variety of channels—calling officials to account, speaking to foreign journalists and expressing their anger on homemade banners. As we have argued, issues of blame and accountability are central to disaster coverage, and they are often raised through discourses of anger. Nonetheless, there are significant variations in terms of who is allowed to raise these issues and in which ways. These variations, in turn, inform the political consequences of each disaster.

Public Discourse of Anger in British Disasters

To better understand the political implications of anger, we examine two high-profile British disasters: the Aberfan landslide disaster, which occurred on October 21, 1966, and the Paddington rail crash (also sometimes referred to as the Ladbroke Grove crash), which took place on October 5, 1999. We trace changing journalistic practices around the construction of citizens and their emotional expression in the public sphere. We examined all coverage of these disasters for two weeks after each event in two national newspapers, the quality paper, *The Times,* and the mid-market, right-leaning *Daily Mail,* as well as local papers from the area where the disaster occurred: The *South Wales Echo* (Aberfan) and the *Evening Standard* (Paddington). The analysis focused on how anger is articulated, mobilized and managed. In particular, we explored the following questions: Who is authorized to express anger and under what circumstances? What are the subject positions and power relations produced and legitimized by these representations? How is anger related to social processes and concepts such as justice, responsibility and blame? And how are these processes and concepts invoked as the basis of critique of society and politics? Through the process of authorizing individuals involved in the disasters to express their anger in a politicized manner, we suggest, journalism practices may open up a space for the exercise of citizenship. As we have noted, opportunities for explicit political critique offered to ordinary people are, in fact, extremely uncommon in routine news coverage, where most citizen expressions remain largely depoliticized and reactive (Lewis, Inthorn and Wahl-Jorgensen, 2005).

The 1966 Aberfan disaster horrified the British nation and has remained etched on the memories of the people of South Wales. A colliery waste tip collapsed into the mining village of Aberfan and engulfed a junior school, part of an adjacent senior school, and several houses. One hundred and forty-four people were killed, 116 of whom were children. The disaster generated headlines across the world. On the evening of the disaster, BBC's newsreader Cliff Michelmore announced the news about Aberfan in tears, stating the following: "Never in my life have I seen anything like this. I hope I shall never ever see anything like it again. For years, of course, the miners have been used to disasters. Today for the first time in history the roll call was called in the street. It was the miners' children" (cited in McLean, 1999). An inquiry that reported 10 months later found the National Coal Board (NCB) fully responsible for the disaster. It concluded that the disaster was a "terrifying tale of bungling ineptitude" and that the postdisaster behavior of the NCB was "nothing short of audacious" (McLean, 2007). However, no legal action was taken. For many, Aberfan represented the incessant exploitation of miners and the failure of the coal industry's nationalization. As one bereaved husband and parent stated, "I was tormented by the fact that the people I was seeking justice from were my people—a Labour government, a Labour council, a Labour-nationalised Coal Board" (cited in McLean & Johnes, 2000, p. 61) In the 1999 Paddington crash, 31 people were killed and about 400 injured at Ladbroke Grove, two miles outside London's Paddington Station. The report of the public inquiry provided damning evidence of how the companies operating Britain's trains since the privatization of British Rail had consistently placed profit before public safety. The immediate cause of the disaster was an error on the part of the train driver, who had passed through red signals. However, it was not discussed in terms of individual fault but framed in the context of previous train disasters. As such, it offered a way to discuss fundamental problems around rail safety.

Anger was prominently present in the coverage of both disasters, and it was most frequently expressed by citizens (or their representatives, such as a village priest or a union leader). The discourse of anger, then, provided an opportunity for ordinary people to express criticism. Journalists also frequently expressed anger, either disguised as a description of "public mood" or through their own opinions. The use of inferences about public mood as a journalistic strategy for channelling opinion is often seen by journalists as a way of meeting their social responsibility of representing the public and holding the powerful accountable, even if it is only rarely based on substantive evidence of public opinion (Lewis et al., 2005). By contrast, elites expressed anger only when they denied responsibility in response to the assignment of blame. This is in line with Ost's (2004) claim that elites prefer calm discussion since it maintains the status quo; unlike outsiders, members of the elite do not need to employ angry speech to get what they want.

In their Aberfan coverage, national papers offered a relatively restrained approach to emotional expression. The focus was not on the question of how the victims, eyewitnesses or the wider public felt—as it would be in the 1990s—but on the question of "what happened." Anger was communicated through the use of direct quotes from the people of Aberfan. Moreover, it was indirectly communicated in reports of numerous protests and petitions. For example, parents in Aberfan reportedly refused to send their children to temporary classrooms built in the shadow of a coal tip (*Daily Mail,* October 31, 1966). First of all, this approach of representing anger by reporting the statements of those affected by the disaster allowed citizens to express themselves in their own words. Second, by drawing on the practices of the "strategic ritual" of objectivity (Tuchman, 1972) in using sources to make political claims, it left the journalists as impartial observers of the events. This is demonstrated in the example of the *Daily Mail* story about the opening of the inquest:

> Anger erupted from grief at the inquest on some of the Aberfan victims yesterday. One father demanded that his son's death should record: "Buried alive by the National Coal Board." Amid uproar, a mother cried: "He's right. They killed our children."...The man who accused the coal board is aircraft worker Mr. John Collins, 40, who lived next to the school. He was at work when the tip wrecked his home. His wife and two sons were killed. A third son is missing. Mr. Collins's interruption set off series of shouts from among 60 parents. "*Those are the feelings of all of us. We want those words on the death certificates.*" "*Our children have been murdered.*" (*Daily Mail,* "Coal Board killed them, say parents," October 25, 1966, *Daily Mail*'s emphasis)

Here, bereaved parents were entitled to speak up and express anger against the Coal Board, raising issues of blame and accountability on the basis of their loss. Unlike the national papers, the local newspaper, the *South Wales Echo,* registered the prevailing mood in the devastated valley more widely in two lengthy articles summarizing readers' letters that had reportedly been pouring in, implicitly drawing on these letters as a "public opinion thermometer" (Sigelman & Walkosz, 1992). Thus, letters to the editor appeared as the most significant venue for the explicit expression of "seething anger." Here are two passages from the readers:

> I also feel cold anger and resentment that not only are these obscene coal tips raping the Welsh countryside, but that one has now been allowed to brutally cut short the lives of so many Welsh schoolchildren. ("Aberfan: What *Echo* readers feel," October 16, 1966)

> I have bitterness in my heart not only at the dreadful loss of life but at the callous attitude taken by Lord Robens, chairman of the Coal Board. He blames the Press and television for the way they brought this calamity into our homes. What did he want them to do, lift the carpet and brush it under? ("How long will the nation remember?" October 27, 1966)

By contrast, in the coverage of the Paddington crash some 33 years later, national newspapers gave citizens a wider opportunity to express their feelings of anger and frustration. As we saw in the case of South Asian tsunami reporting, the circle of emotional witnesses had expanded to include those who were not present at the event or directly affected by it. Readers, from their subject positions as British citizens, were entitled to express their feelings and opinions on the disaster. These are typical "how something like this can happen" comments which the newspapers collected from angry commuters or people otherwise connected to the more recent disaster:

> As the day drew on more anger did begin to surface as information failed to come forward. Back at the station, 21-year-old Lea McMahon said her best friend's mother was lying in hospital after the crash. "This is the main route into London," she said. "Hundreds of people travel from here every day. How could this happen nowadays? We can do anything, we can send astronauts into space but we can't control safety on our trains." (*Daily Mail,* "A candle for the daddies who will not be coming home," October 8, 1999)

In the Paddington case, the journalists did significant "anger work," generating and managing it, thus going beyond reporting on "how people felt." As the following example from a *Daily Mail* editorial illustrates, anger became an expected and appropriate response to the disaster: "If, as seems likely, the Ladbroke Grove tragedy was caused by an overshoot, it will fuel anger against both the Government and the previous Tory administration" (*Daily Mail,* "Did driver jump a red light?" October 6, 1999). Newspapers, and the *Daily Mail* in particular, acted in a representative role to communicate the prevailing emotions of the (national) community, focussing on how people felt and evoking emotions in readers with highly charged stories, as in the following example, describing a floral tribute (usually communicating grief and loss) expressing anger: "Besides the bouquet was a bunch of lilies with an unsigned message saying: 'Bureaucratic incompetence has claimed more lives yet again, why?'" (*Daily Mail,* "A vision of hell in carriage H," October 7, 1999).

Certainly it seems that conventional power hierarchies in media coverage are reversed when citizens' emotional responses are privileged. For the most part, elites were unlikely to express anger in disaster stories, whereas the reactions of those affected by the disasters were foregrounded and represented as newsworthy. Overall, then, politicized anger at institutional neglect and malfeasance is given voice in disaster coverage, though most prominently through genres marked off as distinct from "objective" news coverage, such as letters to the editor.

Anger needs an object, someone who can be held responsible and toward whom the subject can direct his or her emotions. As mentioned earlier, questions of fault and responsibility are central to disaster news, but they depend greatly on the context and whether the way in which they are raised involves anger. It makes a difference, on the one hand, whether the culpability is clear, and on the other hand, whether those responsible for the disaster are individuals or impersonal social systems, such as a private business or a national industry (and, by implication, the government). The vast majority of expressions of anger was directed at industry, political authorities or the media.

In the Aberfan case, journalists were careful to avoid the apportioning of blame toward the NCB or the government. For example, *The Times'* report on the day after the disaster began with very understated language: "The Aberfan slag heap tragedy points to the contrast in mine safety above and below ground. From the pithead downward, miners say, they cannot move a pace without encountering a safety rule; but on the surface the precautions are less stringent" ("Warning in 1964 about Aberfan threat," October 22, 1966). In the fifth paragraph we were told that "the people of Aberfan were *bitter* yesterday that something had not been done about the tip years ago" (emphasis ours). So what we see is a report that lacks directness or overt blame-seeking. From today's perspective at least, this coverage was subtle and restrained and did not significantly scrutinize the authorities. Indeed, the front-page reports of *The Times* focused on the NCB's efforts at deflecting the blame and presenting the disaster as the product of a possible freak accident. In the comments of the people of Aberfan, however, the blame was squarely placed on the NCB. As the village priest was reported to have said, what happened in the village was "a direct consequence of man's neglect and man's failure to act when every intelligent person must have foreseen a disaster of this kind" (*Daily Mail,* October 10,

1966). The board had reportedly ignored warnings of danger for many years: "Yesterday, Mr. Bernard Chamberlain, of Pantglas Road, Aberfan, said he had been battling for two years to get something done. 'It was only two days ago that I made a further protest against nothing being done'" (*Daily Mail,* "Warning of danger given 3 years ago," October 22, 1966).

In the Paddington case, references in coverage to the public feeling were used to legitimize journalists' own criticism toward the rail industry and the government. The Paddington disaster has been taken as an example of "therapy news" focused on the grief of victims (Mayes, 2000), but in fact, reporting on rail safety easily outnumbered the victim stories. Anger was represented as the general public mood:

> It is there too, in the messages with the flowers, a focused anger that goes beyond the despairing at fate that is the normal response. The thoughts of those who have left flowers at the public site, which is against the wall at Sainsbury's have a theme. "This is crime," one card says. "Forget behaving with dignity and sensitivity. We need and we will scream and shout for justice." (*The Times,* "Quiet voice that says this horror is different," October 10, 1999)

Editorials and comments presented harsh criticism targeted at Britain's privatized rail industry and the government's inability to address safety problems. The use of the editorial as a way for journalists to express anger reflects the unique place of this genre within the culture of journalism. As McNair (2000, p. 31) has pointed out, through opinion journalism, "the institutions of the press take the lead in establishing the dominant interpretative frameworks within which ongoing political events are made sense of." Here is an example from the editorial of the *Evening Standard:*

> Today there is real rage among rail users, who pay some of the world's highest fares for poor service, unpunctual and often filthy trains and also—it seems—the risk of death or injury.— But Mr. Tony Blair should recognize that public anger about the state of Britain's transport system is running high, and will be intensified by what has happened in Paddington. The travelling public cares nothing for dogma and political posturing. It wants answers, and solutions. (*The Evening Standard,* "Carnage at Paddington," October 5, 1999)

Editorials such as these spoke on behalf of readers and citizens, acting as advocates for the public and adversaries to government by expressing the imagined consensus of politicized anger and calling for answers and solutions.

It is apparent, then, that disaster coverage, at least in a national case such as the ones studied here, has the potential to open up larger structural and systemic critiques from the bottom up precisely because issues relating to the apportioning of blame and accountability tend to be *political;* they turn into questions of collective and usually elite responsibility, rather than taking a more common—and depoliticizing—approach of apportioning blame to individuals. For the most part, the discussion of the object of anger unusually allows those affected by the disaster to serve as the "primary definers" (Hall, Critcher, Jefferson, Clarke, & Roberts, 1977) of the story of blame and accountability—their version of events and their explicitly political critique create the framework within which subsequent actors and voices must position themselves. As such, the subject position of the *victim* allows for the making of a politically engaged *citizen* entitled to raise significant questions that require a response.

Conclusion

This chapter has sought to shed light on the ways in which anger is expressed and politicized in disaster coverage. While some disasters are staged as moments of national integration and/or the pursuit of political and corporate projects of control, in some disasters national news media give vent to criticism and allow the angry, contending voices to be heard. This chapter revealed that there are systematic patterns, born out of the conventions of journalism, structuring who is authorized to express anger and under what circumstances and constraints. The subject position of the victim or the individual otherwise touched by the disaster allows for the articulation of anger, either through letters to the editor or in stories where ordinary people were used as sources. As such, it opens up a rare space for "ordinary people" to direct criticism at power holders in society—corporations, governments and other social institutions that they otherwise rarely have an opportunity to directly hold to account. More profoundly, it allows for a democratized systemic critique of complex social processes such as privatization that generally lie outside the bounds of acceptable topics for public debate. This is not to claim that emotions and experiences of individuals are necessarily used in news reports to generate challenges or questions to social institutions. Journalism can also confine its horizons to the private experience, reducing the political importance of emotional expressions.

We have also indicated some historical change in the nature of disaster coverage. While the expression of emotion was usually confined to editorials and letters in the 1960s, the more recent Paddington disaster provides evidence of an increase in opinion journalism interpreting events and emotions and offering views, although sometimes this is done in the name of the public. By the 1990s, news was increasingly mediated through individual experiences and accounts. In coverage of the Paddington rail disaster, journalists developed moral narratives around anger, thus abandoning their conventional position of neutral observers, which structured reporting practices in earlier decades. This supports the claim that journalists have become more active participants in political and social processes, especially in the contexts of social crisis (e.g., Cottle, 2006a; Liebes, Kampf, & Blum-Kulka, 2008). Their agency appears to include individualized emotional expressions, which enables them to engage in explicitly political discursive practices, drawing on the register of anger identified here, to raise structural questions of collective significance which might previously have been outside the bounds of "objective" journalism.

9 Transformations in Disaster Visibility

As new technologies of communication and surveillance proliferate, so images of disaster around the world pervade everyday lives, and bearing witness becomes increasingly mediated. We inhabit a world in which new technologies of surveillance including high-resolution satellite monitoring can now map disasters and record and transmit images of many of them as they unfold over time—whether the destructive path of major storms and fires, the scale of floods and droughts, the erosion of melting glaciers and destruction of native forests or the flows of refugees in humanitarian disasters and the deliberate bombing of civilian safe havens. Mobile telephony and new social media now also grant ordinary people caught up in disasters unprecedented opportunities to directly intervene into the communications environment, sending information and images to family, friends and to the wider outside world. In doing so, they can provide insider and sometimes graphic accounts—whether text messages sent from trapped survivors to families and rescuers in the Haiti earthquake in 2010 or phone video sequences of the earthquake and tsunami recorded and uploaded by eyewitnesses in Japan in 2011. And sometimes those caught up in disasters can bypass traditional means of communication or enter into them challenging or contradicting official claims (Friedman, 2011), as in the July, 7, 2005, bombings in London, when eyewitnesses confirmed early on that suicide bombers and not a power surge, as first suggested by officials, were responsible for the devastation in the capital city's underground train network (Gowing, 2009).

New communication technologies have become deeply implicated in what John Thompson has termed "the transformation of visibility" (1995, 2006), a phrase he coined to emphasize the ways in which the gaze of the modern media renders political elites and power-holders vulnerable to the court of public opinion, especially in respect to gaffes, scandals and other forms of loss of control. "In this new world of mediated visibility," he said, "the making visible of actions and events is not just the outcome of leakage in systems of communication and information flow that are increasingly difficult to control: it is also an explicit strategy of individuals who know very well that mediated visibility can be a weapon in the struggles they wage in their day-to-day lives" (Thompson, 2006, p. 31). As discussed earlier, disasters can indeed sometimes prove morally hazardous if not toxic for political elites and authorities when exposed to increased public scrutiny and potential censure in such moments. Mobile telephony and new social media are now reconfiguring communications power in times of disaster. These new technologies and their rapid uptake around much of the world help to communicatively enfranchise those who were formerly invisible or voiceless in the media or who were at best granted a relatively minor role in the overarching narratives and discourses of mainstream media. Today, following a major disaster, communication flows can circulate widely, even globally, flooding mainstream media as well as alternative communication channels with dramatic scenes, visceral images and emotional accounts. This cacophony of voices and images can also include those with strong views on the causes, consequences and culprits of disaster.

Thompson's social theory of the transformation of visibility is not confined, however, to the immediate political struggles of elites and their challengers in this changing environment of "mediated publicness"; it also extends to consideration of an ethics of global responsibility. Toward the end of *The Media and Modernity* (1995), Thompson reflects on how various media of communication "have helped to create a sense of responsibility which is not restricted to localized communities, but which is shared on a much wider scale. They have helped to set in motion a certain 'democratization of responsibility,' in the sense that a concern for distant others becomes an increasing part of the daily lives of more and more individuals" (p. 263). This concern with the contemporary media and its possible contribution to a new global awareness, cosmopolitan sensibility and ethics of care has been pursued by a number of authors (Beck, 2006, 2009; Chouliaraki, 2006, 2010; Robertson 2010; Silverstone, 2007; Stevenson, 1999) and has been elaborated across preceding chapters (see Chapters 5, 6 and 7). New forms of disaster imaging and mapping facilitated by satellite technology are now also contributing to the transformation of visibility in respect of disasters. Though often no less political or contested when the disaster in question involves political authorities denying the reality or scale of, say, a "natural" disaster in their political jurisdiction, or indeed their own responsibility for causing the human suffering involved, these new forms of disaster imaging, as we shall discuss, are capable of powerfully visualizing and documenting and thereby helping to "bear witness" to humanitarian disasters. Such images increasingly lend further support to Thompson's empirical claims and normative hopes for the role of modern media in the "democratization of responsibility."

This chapter, building on earlier discussions, sets out to address some of these principal transformations in disaster visibility and communication based on new communication technologies and explores their interpenetrations within and across the contemporary communications field. Drawing on accounts of news professionals, the chapter first addresses how new communication technologies are becoming *infused* inside mainstream news production and how news producers have sought to incorporate social media and so-called "user-generated content" (UGC) into their disaster packages and reports.[1] The discussion then explores the rapid rise and involvement of mobile telephony and social media in recent disasters and how and to what extent these are contributing to the reconfiguration of relations of communication power. Finally, we consider new forms of disaster surveillance and how imaging and mapping technologies can powerfully visualize the human and environmental impacts of disasters and how, in turn, these have become powerful assets in the international field of humanitarian disasters and human rights activism. Across these three discussions of reconfiguration and change in the contemporary field of disaster communication, the chapter advances three basic propositions. First, it underlines the importance of empirically attending to the practices of the producers and participants involved in communicating and visualizing disasters if we are to better understand the dynamics and negotiations at work and how these condition disaster communications. Second, it points to the rapidity of change within the field of disaster communications and how major recent disasters provide new insights into these processes of communication change, revealing not only the shaping but also shifting relations of communication power within them. And third, the discussion points to how disaster communications are invariably not best approached in terms of generalizing theoretical approaches wedded to either conceptual ideas and models of traditional media systems on the one hand or new communication networks on the other. In practice, as we document and consider, disaster communications increasingly involve overlapping communication flows and interpenetrating communication forms and also often generate new communication hybrids that traverse both "old" and "new" media. This renders somewhat problematic the tendency to conceptualize the contemporary media and communications field, and disaster communications within this, in essentially dualistic terms, or indeed informed by a notion of progressive supersession of "old" media by "new" media. There is, in fact, considerable flux in the field of contemporary disaster communications, but it is not always helpful, we find, to conceive of this in terms of a communications dualism of "old" and "new" media given the imbrications of both in today's more complex, fast-moving media ecology.

Disaster News and New Communications: A New Interplay?

The advent of new communication technologies in the production and circulation of broadcast news in recent years has undoubtedly influenced the nature of disaster reporting and in ways that emphasize established news organizations' predilections and professional views of what constitutes "good" disaster story reporting, as discussed previously. Broadcast journalists and correspondents are keenly aware of, if not transfixed by, new technologies, understandably perhaps, given their daily dependence on them. They often speak of them in deterministic terms:

> I think technology is the key. I think technology keeps changing everything and the way that we report disasters is changing. The changes are driven by technology. Nobody says, "Right, from next month we're going to report disasters in a different way," somebody just gives you a different piece of kit and then, once you've got that different bit of kit, it makes you do things in a different way. If there was an earthquake in China tomorrow, we would turn up at the scene and one of the producers would start asking around if people had videoed the moment of the earthquake on their mobile phones—because everyone has a video mobile phone these days. We wouldn't have been asking people for their own video five years ago, but that would now be seen as a priority, as would being live as opposed to covering the material for a package. (BBC Shanghai correspondent)

The professional tendency to separate out new communication technologies from the organizational goals and cultural outlooks of news production, granting each new technology a seeming determinism and autonomy, is evident in such accounts and in fact precedes the latest wave of technological advances (Cottle & Ashton, 1999). As we hear in the correspondent's reflection above, the deliberate pursuit of available video of disastrous events with an emphasis on reporting "live," rather than constructing news packages, speaks to some established registers of disaster reporting, and not others, and though certainly facilitated by new technologies cannot be reduced to them. There is no doubting, however, the general enthusiasm voiced by many correspondents in the field for the latest journalism kit, especially when it helps them enhance their professional autonomy in the traditional team approach to producing disaster news.

> Well, it's night and day. I've been a journalist for fifteen years and it's just changed so dramatically and it continues to do so....I shoot and edit my own stuff and, you know, I can literally go somewhere on my own, completely alone and take a small satellite dish, camera and a laptop and be on the air to whoever needs me if one of them has a good internet signal—or even not, because I can use the satellite. All you need is a laptop and a camera and you can literally broadcast radio, TV or on-line from any rooftop or any kind of open space, no problem as long as you're safe. And it wasn't like that even 5 years ago. This has really transformed what we do and I know now there are applications for the i-phone 4 which enables journalists to point the phone at themselves and do live shots down through a mobile phone, which is incredible. (BBC North America correspondent)

The rapid evolution of news technology is not entirely frictionless, however, as it can also increase the possibilities for different ways of reporting. In the pressurized time constraints of current news production, such choices can be encountered as professional dilemmas:

> This technology can also create the most dilemmas and difficulties. Most of the experienced journalists are pretty experienced in making those difficult decisions about what seems appropriate or inappropriate to show, or the tone of language to use, and we're quite comfortable with making those decisions. What is much harder these days, is the kind of "Ok, I've just arrived here—the light's fading and we've got an hour left—should we be setting up a live position, should we be shooting material which can be edited into a package, should we be shooting the "walk and talk"? Technology gives you opportunities but it also gives you dilemmas. (BBC Shanghai correspondent)

How these dilemmas are resolved in practice, as suggested, is not entirely reducible to the technology itself but involves the professional thinking and organizational predilections of disaster reporting. New technologies, then, have entered into the professional world of news production, and, as they have done so, they service different news aims, professional practices and identities. The revolution in new social media also proves no less influential in terms of granting increased emphasis to the producers' pursuit and valorization of news immediacy, dramatic pictures, personalized stories and experiential accounts. Here an assistant editor working for the BBC's Interactivity and Social Media Development initiative reflects on the rise of social media in BBC news:

> In any big breaking story, like a disaster, it's social media and the web that provides us with an immediate channel to get images and stories out, where as in the olden days with conventional media it would have taken hours if not days to get information and picture content out. Now we can get that stuff out almost instantly, within minutes frankly. It allows us to get in touch with real people telling their own real stories in a really kind of powerful and engaged way, in a way that we could never do before. It informs and helps guide the story we're telling as well, because while we're aware that not everyone has access to the web and has the ability to use it to tell their story, it certainly gives us a light into a constituency within the story that we might otherwise not have had. So it gives us information about what is happening on the ground that we can then shoot back to our own reporters and journalists who are in the field and help guide their editorial activities and thought processes. (Assistant Editor, BBC Interactivity and Social Media Development)

As the assistant editor indicates, a high institutional value is now placed on social media in news production and storytelling, with the latter's capacity to enhance news immediacy, audience engagement and sensitizing feedback for journalists all recognized and put to work (see also Newman, 2009; Sambrook, 2010; Wardle & Williams, 2008). It is perhaps the *experiential and emotional realism* communicated by social media and UGC, however, that chimes most strongly with news organization disaster reporting (e.g., Pantti & Bakker, 2009; Williams et al., 2011). Facilitated by mobile telephony and social media, the front-line of conflicts, disasters and other human traumas is now, it is said, brought directly into newsrooms. It is in the newsroom and monitoring services back at the base, that editors and journalists increasingly confront raw, unedited and often graphic images sent in by those who are emotionally caught up in scenes of human destruction. The assistant editor of BBC Interactivity and Social Media Development explains as follows:

> [I]t is the personal testimony provided by UGC that gives the emotional power to the storytelling—unlike much of the professionally shot material, which is one step removed from the events portrayed. It is an emotional power that has an impact on our audience and newsroom journalists alike. But it is not just the graphic images that make up this new frontline. Technology—from mobile phones to Skype—now allows participants and bystanders to share their experiences direct and unmediated. (Eltringham, 2011)

The professional exposure to scenes of human misery and destruction is no longer confined, therefore, to those reporters and correspondents "embedded" in the field of disasters but extends virtually to those back in the newsroom and its monitoring services. The profusion of incoming "raw" materials, including disturbing and violent images from disaster and conflict zones, has become more widely available and, subject to prevailing editorial guidelines, is now available to inform the production of news reports and packages including their journalistic inscription with an often barely concealed injunction to care (see Chapter 5).

New communication technologies and the infusion of social media into disaster communication are not, of course, confined to the world of broadcast news. The rise of digital communications and social media also impacts on the practices of photojournalists and the production of images of disaster for the mainstream press. Here an experienced photojournalist who has worked for many of the world's most prestigious newspapers, including *The New York Times, New York Post, The Washington Post, Le Monde, The Boston Globe, The Guardian, The Independent* and *Scotland on Sunday,* summarizes the increased pressures that the changing communications environment has brought about:

> There are many pressures on established photojournalists at the moment. We are in a moment of great technological change. The tools of production have been democratized to a level never before seen. The speed with which organizations (NGOS and media) are expected to report has increased dramatically; while the money available to pay people to gather, analyze, edit and present the information has been in a steady state of decline. The pressure is on for editors and publishers to present material quickly and with minimal cost. Add to this the fact that information can so quickly move from one platform to another freely, and mix in the high volume of materials created by amateurs, aid workers, and the victims or those proximate to a disaster, there is now much less room and much less value for what the professional can create. But people still yearn for quality information, well presented and selected for maximum impact. We have to find new ways to get our job done and to prove the value of our work. We have to fit into the new flows of information while strongly advocating and proving the efficacy of our trade. (Photojournalist)

Notwithstanding the pressures rooted in the political economy of competitive news production, the photojournalist also discerns progressive possibilities in the new communications environment, possibilities that may help to counteract the reliance of mainstream news organizations on a seemingly narrow range of cultural templates and victim stereotypes. The latter too often distance the human agency and coping strategies of survivors as well as their efforts to resurrect some semblance of normality and everyday life amidst the chaos. A photojournalist elaborates revealingly on the problems involved:

> Newspapers and magazines are often very sensationalist in their selection and presentation of humanitarian emergencies. I was asked by one newspaper to photograph Somalis hiding in trees from crocodiles during a flood. They had heard reports of this happening and asked me to make pictures of it. It wasn't a clearly thought out assignment. How would I get to that one tree in a flooded desert? Drive? Boat? And if I wasn't in the tree with the

> person, how could I protect myself from the croc? What I was able to see on that assignment were people struggling to maintain a sense of normalcy despite the flood. There was a barber who was shaving a customer in waist deep water, a brilliant illustration of people's determination to not be derailed during chaos. The image that ended up being used was of a family forced to move from the flood zone using a donkey cart. I would have preferred to see the haircut image in the paper but the conventions for images of refugees is one of people in flight so the editor chose an image that was most like the images we are accustomed to seeing. (Photojournalist)

Developing on such past experiences and in ways remarkably similar to the accounts of the broadcasters elicited earlier in Chapter 5, the photojournalist declares his principal aim as producing pictures that will encourage audiences to "engage" while simultaneously challenging established conventions.

> I think good reporting and good photography challenges people's assumptions. In hemming too close to the conventions of disaster reporting we exhaust our audience and let them off the hook too easily. Each humanitarian disaster and each individual affected is something new. The narrative approach needs to be shifted a bit away from conventions so as to challenge people to engage. (Photojournalist)

The World Wide Web is also subject to the forces of competition and commercialism that contribute to the ever-increasing acceleration of news production. It nonetheless provides, according to this insider account, a much broader canvas for, potentially, a much richer spectrum of disaster and survivor images:

> Two things are happening simultaneously—the news cycle has gotten faster in recent years so there is more pressure to quickly update websites. At the same time the web is an endless canvas, much more expansive than a newspaper, so there is room for more images. This seems like a great opportunity to offer more complex and nuanced reports while at the same time there is a competing demand to quickly fill space and move on to the next story. (Photojournalist)

In such ways the World Wide Web, though no automatic panacea for rectifying standardized, unimaginative and reductive disaster images, is said to open up new possibilities nonetheless. (The reader may be interested to follow up on some of the work of the photojournalist concerned and his efforts to provide new images of those embroiled in humanitarian disasters at http://www.brendanbannon.com/gallery-list.)

In the previous accounts, the transformation of visibility within the mainstream media and facilitated through new means of communication has served, to some degree at least, to communicatively enfranchise those who were formerly subject to, rather than agents of, media and communication. This reconfiguration of relations of communication power, however, is unevenly distributed across the communications field. Within the BBC, for example, arguably a world bastion of journalist notions of objectivity, impartiality and detached and balanced reporting, new media technologies are exacting subtle shifts in the traditionally settled configurations of communication power but not radically changing them.

> The relationship between mainstream media and its audience, I think, is shifting slowly and it's a real tension between journalists who are used to controlling the message and everything about the message, and now that control is being prized away from them a little bit. Still, I think certainly the BBC very much do control what goes on our output. We look at the content across the social web and decide whether it's relevant, valid or authentic and we then add it to the global mix and make our own editorial decisions, informed by these new voices. We maintain overall editorial control but I think what the social web does is that it increases transparency and accountability and makes us as journalists a bit more honest frankly. Because there is always somebody out there on the web who knows more than you about a story....So, coming to terms with all of that is quite hard and will take a bit of time, but I can see it happening already and it's hard for a lot of people within the organization to come to terms with. But that's what's interesting and exciting about this area. (Assistant Editor, BBC Interactivity and Social Media Development)

The incorporation of social media has, according to such insider accounts, undoubtedly played an important part in subtly reconfiguring relations of communication power in the BBC's news reporting, but this does not imply, of course, that the BBC has relaxed its proprietorial hold on processes of editorial control and the power to decide who is permitted to speak and what is shown under its watch (see also Newman, 2009; Sambrook, 2010; Wardle & Williams, 2008).

The discussion so far has discerned how new communications in the field of mainstream news production and disaster reporting have been incorporated into journalist and reporting practices and granted increased emphasis on, for example, immediacy and live reports, the pursuit of dramatic images and emotional eyewitness accounts. This is not simply the outcome of a technological determinism, though journalists are apt to portray it in such terms; rather, it reflects how new communications technologies are "put to work" within the competitive, pressurized and culturally inflected regimes of established news organizations. The transformations of visibility brought about within these technologically facilitated forms of disaster reporting can prove, nonetheless, no less significant for that and resonate emotionally with audiences and producers alike. But perhaps the key point here is that the reins of editorial control and the crafting of disaster stories remain firmly in the hands of the mainstream news producers and notwithstanding their increased exposure to an expanded array of views and voices emanating from the heart of disasters. If this suggests that mainstream news media have generally incorporated and creatively adapted to the changing communications environment including the rise of new media, how do new social media enter into disaster communications more directly and with what impacts?

Disasters and Social Media: Shifting Communication Power?

Noticeably across recent years, major disasters have often been associated with the latest communication technologies, an index in part of the rapid pace of technological change now characterizing the wider media and communications environment. Some of the most dramatic images of 9/11, now fixed in historical memory, were captured on personal camcorders, and the voices of many of those trapped in the twin towers became immortalized in messages sent on cell phones and left on answering machines for their families and loved ones. The Asian tsunami of 2004 also became recorded for posterity on video cameras, filmed by holidaymakers from hotel balconies where the audible gasps of disbelief and then panic could be heard over the incoming surge of water. In Haiti in 2010, the role of mobile phones and text messages sent by earthquake victims caught in the rubble captivated news and wider interest. And in the Japanese earthquake, tsunami and nuclear meltdown at Fukushima in 2011, video phones, Skype and various forms of social media all played a prominent part, as did stunning live coverage from NHK, Japan's national broadcaster, who put helicopters into the air and broadcast live, as it happened, the relentless waves surging inland as the environmental and human devastation piled up along the Pacific coast. Our collective memory of major disasters, it seems, has become dependent on and thereby in important respects defined by the latest means of communication. Fast-evolving personal communication technologies, including mobile telephony and social media, are now providing an independent means of recording and relaying images of disaster, often based on eyewitness accounts (Allan, 2006; Allan & Thorsen, 2009; Chouliaraki, 2010; Friedman, 2011; Zelizer, 2007).

It is instructive, therefore, to consider exactly how these forms of personal and social media have recently informed disaster communications, transforming their visibility and possibly reconfiguring relations of communication power. The disaster relief community generally sees the devastating earthquake that hit Haiti on January 12, 2010, with an estimated loss of 316,000 lives, with countless more maimed and at least 2 million homeless out of a population of 9 million, as something of a communications turning point. According to the report, *Media, Information Systems and Communities: Lessons from Haiti,* the Haiti disaster became a laboratory for the application and development of new communication technologies (Nelson, Sigal, & Zambrano, 2011), including SMS (short message service) texting, interactive online maps and radio-cell phone hybrids. By these means, search and rescue teams were directed to people buried under the rubble, missing persons were logged and reunited with surviving family members, and information about sources of food, water and health care publicly disseminated—developments that are anticipated to be considerably expanded in future disasters (Nelson et al., 2011; see also United Nations Foundation, 2011a).

In the context of the catastrophic effects of the earthquake, these communication developments proved all the more impressive. The Haitian government and local media's capacity to respond to the earthquake had been severely undermined by the overwhelming scale of the destruction: 18,000 civil servants had been killed and most public buildings in the capital, Port-au-Prince, had been completely destroyed. The impoverished nature of the Haitian state also meant that it was in no position to deliver effective disaster relief including food, water and heath care on the scale needed. Many journalists had also been killed and Haiti's news media had been effectively put out of action. It should also be noted that the literacy rate in the country at the time was only 52%. Radio had therefore proved to be Haiti's dominant medium before the earthquake, with over 250 commercial and community-based radio stations. Only one managed to keep broadcasting following the earthquake, Signal FM, and this to an audience of almost 3 million, a third of the entire population (Nelson et al., 2011). In fact, radio quickly established itself as a key communication lifeline, providing round-the-clock coverage, helping survivors to locate families and putting out public service messages. Community networks, including local churches, also distributed messages by this means. The importance of radio, however, was considerably expanded when combined with mobile text messaging and social networking.

Though only 10% of the Haitian population could have been described as Internet users at the time of the disaster, in keeping with other poor countries, Haiti had relatively few landlines (108,000 in its population of 9.6 million, ranking it on a per capita basis as 142nd in the world). Nonetheless it also shared in the world's recent exponential rise in mobile telephony; there are now 5 billion phones in the world and 4 billion of these, the vast majority of which are mobile phones not landlines, are in developing regions (Edelstein, 2011; Fox, 2011; Smith, 2009). Haiti had 3.6 million cell phones. This played a significant part in the disaster communications environment. The widespread availability of mobile phones, or access to someone who owned one, facilitated a platform for SMS services and proved to be "a critical interface between the Haitian public and international humanitarian organizations" (Nelson et al., 2011, p. 12).

Numerous Web 2.0 or interactive Internet-based communications, including new hybrid initiatives, were also piloted for the first time in the context of Haiti and are documented in the United Nations Foundation's report, *Disaster Relief 2.0: Information Sharing in Humanitarian Emergencies* (2011a). Here, three examples help indicate how new communication systems are changing the nature of disaster communications and in ways that expand the range of views and voices involved.

The UN Office for the Coordination of Humanitarian Assistance (OCHA) turned to Crisis Mappers and Sahana—examples of the new volunteer and technical communities—to use crowdsourcing to geolocate 105 health facilities in the context of devastated Haiti. Here Internet-savvy volunteers based in different countries rallied to the needs of the disaster victims and rescue services, pooling their time and capability via the Internet to monitor and build an updating picture of the disaster and its destructive effects. By these collective and virtual means they quickly managed to locate 102 of the 105 missing hospitals, inputting all the data into the Sahana disaster management system and verifying it through high-resolution satellite images. These data were then made available in open data formats and became one of the best sources for information about health facilities for the next month. Crowdsourcing had accomplished in just over a day what would normally have taken considerably longer.

Mission 4636, the free emergency communications number made available during the post-disaster period, was coordinated by a team at Stanford University and made use of volunteer and technical communities, facilitating members of the Haitian Diaspora to translate thousands of messages from Creole into English. Creole speakers were recruited through Facebook and other public websites and asked to translate tweets, SMS messages and Facebook posts coming through the free emergency communications number, 4636, from Haiti. During the first month, more than 130,000 text messages went through 4636, and at its peak, 1,200 Haitians were translating thousands of messages per day and usually within minutes of their arrival.

Mobile telephony in Haiti, as we have heard, was extensive, with approximately 85% of Haitian households having access to mobile phones at the time of the earthquake. The cell phone towers were quickly restored after the disaster and facilitated Mission 4636, which enabled both the receiving and sending of SMS messages. According to the United Nations Foundation, "Haitians sent hundreds of thousands of messages out via SMS to Twitter, Facebook, Ushahidi, the Thomson-

Reuters Foundation Emergency Information Service (EIS), and most importantly, to members of the Haitian diaspora" (United Nations Foundation, 2011b). By such means, communications originated from those affected by the disaster as well as those charged with the responsibility to respond.

In these examples, then, we begin to see how new developments in communication technologies became harnessed to specific community needs and applied, innovated and hybridized with other forms of communication media. New "communities" also thereby came into focus: the recently named "volunteer and technical communities." These comprise a variety of technical organizations and voluntary groups from anywhere in the world that share a desire to help and have a virtual (and sometimes on-the-ground) presence, but all based on computing networks and capabilities. This, too, represents a noticeable development in the world of disaster communications and the involvement of wider civil society. While debate continues, for example, about how mainstream images of disaster may or may not communicate feelings of compassion and prompt action, here at least, a desire to help was quickly translated into concrete action and enacted through new media forms of virtual connectivity. Monique Villa, chief executive of the Thomson-Reuters Foundation, which produces the online website AlertNet, underlines the radical nature of this departure when describing the traditional approach to disaster communications and its inherent communication gap:

> When disaster strikes there is a huge gap in the information chain. You have the news organizations from all around the world who come in and cover it, and we do that too at Thomson-Reuters. And then you have the humanitarian machine trying to save lives, but no one is giving the local population information that could be vital to them. That was the gap. (Monique Villa, cited in Bulkley, 2010)

The chairman of the United Nations Foundation, Ted Turner, also commented in *Disaster Relief 2.0* how this "huge gap in the information chain" was addressed in Haiti and how this presages new forms of civil society participation and humanitarian efforts in disasters in the years ahead:

> The global response to the January 2010 7.0 magnitude earthquake in Haiti showed how connected individuals are becoming increasingly central to humanitarian emergency response and recovery. Haitians trapped under rubble used text messaging to send pleas for help. Concerned citizens worldwide engaged in a variety of ways, from sending in donations via SMS, to using shared networks to translate and map requests for assistance. Powered by cloud-, crowd-, and SMS-based technologies, individuals can now engage in disaster response at an unprecedented level. Traditional relief organizations, volunteers, and affected communities alike can, when working together, provide, aggregate and analyze

> information that speeds, targets and improves humanitarian relief. This trend toward communications driven by and centered on people is challenging and changing the nature of humanitarian aid in emergencies. (United Nations Foundation, 2011a, p. 7)

Recent developments in communication technologies alongside civil society initiatives such as these are not confined to the Haiti earthquake. Ushahidi (which means "testimony" in Swahili) has become a widely known online platform since its inception when deployed to map reports of violence following the Kenyan election of 2007. Since then, this open access, crowdsourcing tool has been applied in diverse crisis situations where citizen-generated information is sent via SMS, e-mail, or the World Wide Web and collated and visualized into time-sensitive, interactive geographical maps. "The main goal of the organization," according to the company website, "is to create a system that facilitates early warning systems and helps in data visualization for response and recovery" (http://www.ushahidi.com/about-us). Only a few hours after the Japanese earthquake struck in March 2011, Japanese volunteers, with the help of U.S. academics, had set up their own version of Ushahidi. This helped to identify locations where people could be trapped, areas to be avoided and places where supplies of food and clean water could be found. ABC, the national Australian broadcaster, also deployed Ushahidi in January 2011 to map the floods in Queensland. And Ushahidi helped to map and visualize ongoing developments in the Arab uprisings across North Africa and the Middle East among its many other recent applications (http://www.ushahidi.com/products/ushahidi-platform). Further developments in the communications field include new iPhone apps designed to raise funds quickly for humanitarian relief, alerting diasporic communities around the world, for example, to the humanitarian needs of their own nationals and facilitating instantaneous, electronic transfer donations.

It is unsurprising perhaps that many will see the rise of these new disaster communication developments, involving social media and new communication hybrids, as profoundly empowering for citizens both inside and outside the disaster zone. Charlie Edwards, of Demos, for example, maintained the following:

> Significantly, the difference between the traditional model of communication and social media is the invitation to act. Broadcasting messages offer information, in the form of advice and guidance, to the public. Social media changes that equation by inviting us to take part and share information with others. This is crucial when it comes to disaster management and emergency planning. No longer is the focus solely on the individual but on the individual *and* their network—something that is crucial when important information needs to be passed on swiftly....social media allows individuals and communities to share and cooperate with one another outside of the framework of traditional institutions and organizations. (Edwards, 2009, p. 2)

Those involved with Web 2.0 developments in the field of disaster communications are apt to adopt a more qualified stance of optimism, one informed by the practical difficulties of coordinating communications in contexts such as Haiti, where organizations on the ground exhibited different views on the sharing, control and management of information. The authors of *Media, Information Systems and Communities: Lessons from Haiti,* for example, reflect on these experiences, as follows:

> But while the democratic approach to information management fuels crowdsourcing, this characteristic can also serve as a limitation in crisis settings. Information may be gathered and assembled in an open, democratic fashion. But often the practical response effort is driven by large organizations that deal with information in a radically different way. Military and international humanitarian organizations manage information within more closed systems. (Nelson et al., 2011, p. 9)

Government departments particularly, they go on to observe, function within complex bureaucracies, and these are "accountable to legal frameworks, governing issues of secrecy, privacy and accountability" that make them less flexible than NGOs (Nelson et al., 2011, p. 9). Different relief organizations, including the proliferation of NGOs in recent years, each with their differing remits, stakeholders and cultures of making things happen, inevitably produce frictions. As cyber activists have sometimes complained, "Technology is easy, community is hard" (Nelson et al., 2011, p. 7). In other words, the rise of new social media and new communications in disaster communications cannot be evaluated outside of the relevant organizational field and its composition by different institutions, goals, interests *and* views on how the communications environment needs to operate.

The necessity for fast, efficient communication systems immediately following a major disaster, capable of helping to coordinate multiple levels of national and international governance as well as scores of humanitarian NGOs (Cottle, 2009a, pp. 146–163), has been recognized for some time. Télécoms Sans Frontières (TSF), the leading humanitarian NGO specializing in emergency telecommunications, was set up in 1998 with the express mission facilitating communications in disasters. TSF crews comprising information technology and telecom specialists now intervene anywhere in the world in less than 24 hours, setting up in minutes satellite-based telecom centres offering broadband Internet, phone and fax lines. In 2006, TSF became a partner of the United Nations OCHA and the United Nations Children's Fund (UNICEF) and is generally recognized as a key component of the Emergency Telecommunications Cluster (ETC; http://www.tsfi.org/en). The elevation by OCHA of "Communicating with Disaster Affected Communities" (CDAC), to the position of a lead cluster role in the Haiti earthquake, further underlines the centrality afforded to communications in recent humanitarian operations. But the upsurge in new channels and flows of communication emanating from those directly caught up in disasters can also prove challenging to established disaster relief bodies and NGOs.

> Beneficiaries now have a voice, and affected communities have a voice. They're not just recipients...they have the ability to talk back. That (two-way communication) creates a different dynamic around accountability and responsiveness. It also creates a new set of ethical responsibilities, especially around expectations and whether they can be met....Humanitarian organizations have always prided themselves with being responsive to beneficiaries' needs, and being accountable to them. But there is now a different set of tools to make that happen and it is taking some organizations by surprise. (Katrin Verclas, cited in United Nations Foundation, 2011a, p. 13)

In disasters, the deployment of new forms of communication, as we have heard, can certainly be interpreted as communicatively enfranchising and, to that extent, empowering, opening up new communication pathways and bringing into the field of emergency relief an expanded range of voices, views and expertise. Survivors from the heart of the disaster zone, voluntary and technical communities on the ground and operating from around the world, as well as a plethora of NGOs, tiers of national and international government and of course the world's media, now all enter into an increasingly cacophonous field of disaster communications (Cooper, 2011; Cottle & Nolan, 2007). New communication flows directly from the disaster zone and communicating survivors' needs to humanitarian agencies as well as a flow of images and firsthand accounts to the mainstream media are also challenging the traditional gatekeeping monopoly of NGOs and mainstream news media in initiating and controlling such communicated views and voices. And by these means, an expanded array of views and voices, including those who may sometimes want to challenge the policies and priorities of governments and relief agencies, can also enter into the media public eye.

By definition, however, the field of disasters is not an ideal terrain for an experiment in participatory and deliberative communications, and perhaps it should not be so construed. As we have observed with the incorporation of social media and UGC inside the BBC, though relations of communication power have become expanded to a degree and redrawn by the "civilian surge" (Gowing, 2009) of views and voices both sending and receiving disaster and crisis communications, this does not necessarily translate into a radical restructuring of hierarchical and institutional power. Survivors, whether communicatively enfranchised or not, remain in a situation of extremis and, despite their efforts to help themselves that generally go unreported, remain in a position of extreme precariousness and structural dependency. Such conditions are not conducive for sustained, deliberative political engagement, even when communications become more widely available. Transformations of disaster visibility, in this sense, though important and even communicatively enfranchising to some degree, cannot mitigate the conditions of "bare life" confronting many disaster survivors and that structurally position them in an unequal relationship of powered subordination and dependency. Not only their *life chances* but, existentially, their *chance of life,* remain dependent on the wider knowledge and visualization of their plight. Here, at least, a further recent development in the transformation of disaster visibility also enters into the field of disaster communications and potentially contributes to extending the concern for distant others and the "democratization of responsibility" (Thompson, 1995, p. 263), discussed next.

Remote-Sensing and the Visualization of Humanitarian Disaster

The media's role in "bearing witness" to disasters has formed a continuous theme across many of the preceding chapters and was the explicit focus of Chapter 5, where we explored journalism's seemingly deep-seated antinomy in respect of its entrenched "calculus of death" and emergent "injunction to care." There we heard how disaster journalists and correspondents continue to place significant professional store on "being there" in the field of disaster and witnessing close-up and via all their senses (but sight especially) the scenes of devastation and human destruction. This bodily exposure and "sensing," rooted in the ontology of physical witnessing, then became performatively invoked and deliberately crafted in and through the journalist's narrativization and visualization of disaster news, in its epistemology of witnessing. This last discussion about the transformation of disaster visualization in the contemporary media and communications environment turns to a new technological means of "sensing" disasters, one that also privileges the visual and seeing but one that, contrary to the journalist's bodily immersion within the disaster zone, is conducted remotely. However, as we shall consider, this "remote-sensing" can sometimes perform a no less powerful role in global acts of witnessing when communicated via mainstream media around the world and when actively interpreted through a cultural and political lens of humanitarianism and human rights.

Geospatial technologies, including satellite remote-sensing, now increasingly feature in the mapping and visualization of disasters, including conflict-related humanitarian disasters and human rights abuses around the world. A network of commercial, military and scientific satellites orbits the Earth and combines with powerful geographical information systems (GIS) and Internet-integrated mapping technologies. Together, these dramatically extend the capabilities and social compass of the technical means of satellite surveillance. GIS can generate layered maps that literally "visualize" developments on and below the Earth's surface, rendering such phenomena perceptible and geolocating them with precision in both time and space. They can also map and visualize the course of disasters and destructive human activities with the same precision. The technologies can be used to help predict and locate the scale and severity of natural disasters such as storms, fires, blizzards and floods. They can also chart and visualize the environmental consequences of man-made accidents and disasters such as the marine and coastal devastation caused by the BP oil spillage in 2010. These tools have proved to be a powerful adjunct in the science and communication of (*un*)natural disasters also, such as the changing climate, documenting through time-series images accelerating processes of deforestation and desertification, melting glaciers and rising sea levels, and drying lakes and changing human habitats. They can also powerfully be put to work in the field of humanitarian disasters and human rights advocacy.

The American Association for the Advancement of Science (AAAS), a non-profit organization, is one of many that have expanded the applications of geospatial technologies. Through its Geospatial Technologies and Human Rights Project, it analyzes remotely generated data, providing critical information on the impact of conflicts on isolated communities and environmental and social justice issues and equipping NGOs with rapid authoritative data that can be especially pertinent during times of fast-moving crisis (see http://shr.aaas.org//geotech/ge.shtml).

The Satellite Sentinel Project (SSP), run by the Harvard Humanitarian Initiative at Harvard University, also demonstrates the power of networks as well as sophisticated satellite imaging technology. As we write this chapter, in August 2011, the SSP has produced evidence consistent with on-the-ground allegations that the Sudan Armed Forces (SAF) and its militias have engaged in a campaign of systematic mass killing of civilians in Kadugli, South Kordufan. Based on an analysis of satellite imagery and eyewitness reports, it has identified a site consistent with mass graves in Kadugli. Images such as these prove to be powerful visual assets in documenting, verifying and ultimately prosecuting those responsible for human rights violations. As the SSP website reminds its viewers, "Under the Rome statute and other international humanitarian law, the systematic killing of civilians in peace or war by their own government can constitute crimes against humanity" (http://hhi.harvard.edu/programs-and-research/crisis-mapping-and-early-warning/satellite-sentinel-project).

Notable instances of satellite images confirming human rights violations against international law in recent years include the systematic destruction of villages in Darfur carried out by the Sudanese government and Janjaweed militias and the Sri Lankan government's deliberate bombing of civilian "safe havens" in the final stages of its civil war against the Tamil Tigers, killing over 40,000 people, most of them noncombatant civilians. When circulated in the mass media, images such as these can perform a vital role in galvanizing public and political responses. The satellite images of the Sri Lankan bombing of civilian areas, including the repeated targeting of hospitals, for example, were carried in many national newspapers and by prominent TV broadcasters including Channel 4 news in the UK. Channel 4 also produced a much publicized special investigative programme, "Sri Lanka's Killing Fields," based on a careful time analysis of satellite images and disturbing mobile phone video taken by refugees and so-called "trophy footage" of summary executions filmed by members of the Sri Lankan military.

When circulated more widely in the public domain, these terrible images functioned to "bear witness" to the horrors endured by the refugees and the murder of scores of thousands of Tamil civilians. Channel 4's investigative programme was first shown at the UN Human Rights Council (June 3, 2011) and on Channel 4 (June 14, 2011) and, at the time of writing, remains available on the Channel 4 website (http://www.channel4.com/programmes/sri-lankas-killing-fields). The public and political outrage following this documentation of Sri Lankan military atrocities reverberated in political circles and institutions, including the British parliament, European parliament and United Nations. In such ways, Thompson's "transformation of visibility" brought about by media and communications can arguably begin to contribute to the "democratization of responsibility."

Both the AAAS' Geospatial Technologies and Human Rights Project and Harvard University's SSP provide extensive lists of how their various reports and findings have registered across countless mainstream media reports, effectively documenting their respective organization's reach into wider civil society and dependency on mainstream media. It is only when incorporated and disseminated in the wider flows and communication circuits of the national and international media that these remote satellite images, "visually sensing" human destruction and catastrophic conflicts around the world, can help to undergird today's "global human rights regime," instantiating the normative outlook of global civil society and reinforcing its institutionalized and legal framework (Stanyer & Davidson, 2011). Again, we find that the transformation of visibility in the context of humanitarian disasters and human rights violations is powerfully enacted through fast-evolving communication technologies. As in Haiti, where we learned of how new hybrid communication technologies and interpenetrating communication flows opened up new communication possibilities, so in the case of satellite imaging we find that the cultural and political charge of such images only becomes released in and through the wider communications environment. The new geospatial technologies and their social extension are based on a sophisticated amalgam of satellite remote-sensing, computer information systems, global positioning systems and Internet-integrated information mapping tools, and, in the earlier examples, we have also heard how their wider public and political reverberation depends on interfacing with the mainstream news media.

As we also discussed in the context of disaster communications in Haiti, what we are witnessing here is a far cry from a simple technological determinism. How new communication technologies enter into the existing ecology of media and communications and become deployed in practice and in and through surrounding social and political relations undermines claims that this is a simply a matter of the latest technological developments. Bearing witness in a globalizing media age may well depend on media and communications to help "bring home" the plight of distant others, but how this becomes deployed and communicated in practice and how we feel obligated to those we witness by such means depends on complex processes of mediation, including prevailing sensibilities and extant ethical outlooks. As we just heard, the potential capability of satellite remote-sensing requires a community of human rights activists and a wider normative culture of human rights to put such technological means to progressive purposes and only then, in its social and political enactment, can it further instantiate and help normalize and legitimize this cultural horizon and global movement for political change.

Conclusion

Disaster communications, as we have heard, are fast-changing, transforming disaster visibility around the world and reconfiguring relations of communications power as they become deployed, adapted and incorporated into the complex circuits and overlapping communication flows of today's media ecology. Though in part this comprises new forms of communicative enfranchisement, expanding the range of views and voices emanating from within the zones of disasters and enhancing opportunities for bottom-up and horizontal communications from at least some of those directly involved, the perilous conditions of disaster, by definition, continue to position survivors in often dreadful, life-threatening circumstances, and this precariousness, compounded by relations of structural inequality and dependency, remains politically and humanly emasculating. John Thompson's ethical hopes for the media's role in the "democratization of responsibility," which helped to open this chapter, nonetheless find some evidential support when exploring how new technologies of communication are now being deployed to help visualize the plight of distant others and transforming the visibility of disasters around the world.

In this chapter, we have sought to explore the sometimes complex and differentiated ways in which new media and communications have entered into the field of disaster communications. This includes how new social media have become editorially embraced and incorporated within established mainstream television news and how the World Wide Web has helped to encourage, in an accelerating and increasingly competitive commercial environment, photojournalist creativity and innovation in respect to images of disasters and survivors. Hybrid applications based on crowdsourcing technologies, new social media and mobile-telephony have further proved to be powerful communication tools, enhancing the work of emergency services, communicatively enfranchising some survivors and encouraging new forms of civil society involvement through new tech-savvy voluntary and technical communities—though not without some frictions along the way, given the challenges these pose to established organizational and institutional hierarchies and communication practices. And finally, we observed new forms of "remote-sensing" (or sensory-witnessing) facilitated through the creative fusion of satellites, computing and software that together, when put to work by human-rights activists and others, can accurately map and visualize humanitarian disasters and the abuse of human rights. When such images enter into the public domain via mainstream media and local-global communication circuits, they sometimes reverberate politically around the world.

More than two centuries ago Immanuel Kant wrote the following: "The intercourse, more or less close, which has been everywhere steadily increasing between the nations of the earth, has now extended so enormously that a violation of right in one part of the world is felt all over it" (Kant, 1795/2005, p. 20). As many have commented since, including the authors of this book, the journey to cosmopolitan maturity has proved, it seems, to be a far more enduring and complex affair than Kant may have originally suggested. On the basis of the transformations of disaster visibility facilitated by new communication technologies and discussed earlier, we may now want to conclude the following: "The flows of global communication, which have been everywhere steadily increasing between and beyond the nations of the earth, have now extended so enormously that major disasters and the violation of human rights in one part of the world can often be communicated and witnessed all over it." This chapter has hopefully helped to illuminate something of how the fast-changing media and communications environment is now profoundly implicated from local to global levels in the transformation of disaster visibility around the world and how, notwithstanding the different dimensions and complicating dynamics explored across this study, many of us can now bear witness to them and in ways that encourage the democratization of responsibility.

Note

1. The term *user-generated-content* (UGC) does not accurately encompass the range of new media or their increasingly diverse social, political and cultural deployments. Originated in the industrial and professional world of broadcasting, it suggests perhaps a particularly stunted view of "its" user audience and mainstream media incorporation. It also conflates different processes and motivations for media engagement, and its conceptualization of 'content' simply flattens the complexity of different forms and appeals of media representation. As the continuing evolution of new communications, rise of citizen journalism and deployment of social media in major crises and conflicts, such as the Arab Spring attest (Cottle, 2011b), the technicist and politically evacuated notion of UGC has arguably passed its corporate sell-by date.

10 Conclusion

DISASTERS DRAMATIZE OUR COLLECTIVE VULNERABILITIES AND THE RISKS THAT WE all live under. If anything, these risks and vulnerabilities are only increasing in times of global climate change and population growth. Today, as we have argued, media constitute disasters, shaping how they become understood and responded to. Our knowledge about the fate of distant others suffering the consequences of cataclysmic events, and the spaces and communities that they shatter, is gathered through the media first and foremost. We have traced an increasing sophistication of academic resources for understanding the social, cultural and political implications of disasters. We have made the point that it is only through an understanding of the complexities of globalization and the profound influence of media and communications in their wider signification and subsequent responses that we can trace how global power relations shape our understandings of catastrophic events and their victims—and with what consequences.

With this in mind, we have sought to provide broader insights into how the media constitute disasters and what the consequences are of the systematic ways in which images, texts and discourses around disasters circulate and are made meaningful. Woven through our discussions and examples has been the attempt at problematizing the prevailing idea of "disasters" as unpredicted and calamitous events and instead reconceptualize and resituate them in the context of a rapidly globalizing and crisis-prone world. As disasters become defined and reported in global news media, so their repercussions and responses can outstrip their immediate geographical setting. Moreover, we have made the point throughout this book that disasters are inextricably bound up with wider discourses and reflect larger tensions at global and state levels, insofar as they enact conflicts, solidarities and sympathies and variously enable and foreclose forms of political and social action. In the context of increased global interdependency and corresponding hopes for the emergence of a cosmopolitan imagination, our book has outlined some of the complications, qualifications and possibilities of disasters as politically and socially meaningful events, and this in a time when we can only take their increased presence and force for granted.

In our attempts to grapple with the meaning of disasters, we have taken a particular interest in three closely related conceptual arenas, which help to map out major unfolding debates: those of globalization, citizenship and emotion. We consider these interrelated conceptual arenas as key to understanding the complicated contours of contemporary mediated disasters, not merely because they are central to so much scholarship in the area, but also because they provide us with useful tools for thinking through the relations, discourses, practices and subjects constructed through disaster coverage. This approach has allowed us to explore how disasters dramatize broader themes around—to mention just a few examples—the possibilities for cosmopolitan citizenship to emerge through disaster coverage, the place of "ordinary people" in the newsworld, the role of victims' and journalists' emotions in framing meanings around disaster and the ways in which an emerging globalized world creates unprecedented interconnections between and among individuals, regions, nations and groups. At the same time, the book has uncovered how the relationships fostered by globalization and played out in the coverage of disaster are heavily complicated by power relations and geopolitics, and mediated by changing communication technologies and emotional discourses.

Taken together, the chapters in the book make the case that media coverage of disasters—when viewed as a form of metadiscourse on politics in the nation state and beyond, as well as on our views of proximate and distant others—matters greatly. This is so not just in terms of audience understandings of *particular* disastrous events but also more *broadly* in the cultural elaboration of shared meanings. In this context, our exploration of long-standing and emerging logics around media reporting of disasters at once highlights progressive potentials but also stakes out their limitations. It invites us to understand mediated disasters structured by a series of deep-seated, often unresolved tensions or seeming antimonies—rationally antithetical but in practice muddling through and sometimes in flux. These structuring tensions concern relations between the *global and the local,* the *individual and the collective* (between state and citizens, the macropolitical and the micropolitical), conventional journalistic ideals of *objectivity and impartiality* and an emergent attention to *emotional expression* encoded into vivid stories about the suffering of distant others and journalists' own emotional reflections, and also the professional, epistemological and ultimately political tension between *traditional mainstream media* and fast-evolving *new forms of social media.*

First of all, disasters can be located within contemporary and ever-dynamic media landscapes constituted through the *tension between the global and the local,* as "global focusing events" or "cosmopolitan moments" ever tempered by forms of "banal nationalism" and geopolitically circumscribed interests. As we demonstrated in Chapter 2, "Media and Disasters in a Global Age," disasters can no longer be viewed as solely or even primarily as geographically bounded events that impact only on local and national arenas. Instead, disasters are often distant events which nonetheless can "touch home," recognizing the "other" as not so different from "us" and eliciting global communities under the negative sign of a "world risk society." Earthquakes, environmental devastation and tsunami waves are no respecters of national boundaries, and disasters increasingly have political and economic implications that reverberate around the world. Moreover, disasters are mediated in and through local-global news formations and no less globally dispersed and originated communication flows. Little wonder perhaps that scholars and practitioners alike are increasingly calling for a "global journalism" (e.g., Berglez, 2008); the hard evidence of increasing global interdependency and inequality is often glimpsed in times of disaster and may yet help to sustain an emergent cosmopolitan imagination. Our book has delineated both the opportunities and challenges of such a possible transformation.

The conceptual and empirical framework of the book has often remained focused on the context of national and local media as central to the consumption practices of audiences around the world. Even if such media practices draw centrally on forms of domestication, we have also uncovered the potential for a cosmopolitan imagination to emerge out of national media contexts, and indeed the two need not be seen as antithetical when the principal of *both/and,* not simply *either/or,* is recognized and operative (Beck, 2006; Berglez, 2008; Berglez & Olausson, 2011; Cottle & Lester, 2011). Along those lines, our examination of the coverage of the 2004 Boxing Day tsunami in Chapter 6, "Compassion, Nation and Cosmopolitan Imagination," demonstrated that the coverage successfully facilitated the imagination of others' suffering and, at the same time, a cosmopolitan future based on new emotional bonds with others. Narratives about the caring and generosity of local people and the gratitude of the Western victims used in the national media were particularly compelling. What emerged in the tsunami coverage could be defined as a moment of "internal globalization" (Beck, 2002), a transformation of the public sphere at the national level, as global concerns became part of local experiences and moral life worlds. Representing suffering with a strong national reference, with local people in the afflicted areas featuring as moral agents in the role of collaborators, resulted in a "wave of compassion." On the other hand, news about the relief response forcefully created an active public and invited the readers as citizens to participate in public action.

At the same time, we also demonstrated that disaster coverage and the forms of citizenship it constructs are heavily influenced by geopolitics at the levels of ideology and journalistic practice. As Chapter 3, "The Geopolitics of Disaster Coverage," argued, disaster coverage is inherently political, insofar as it always points to the nation state and its role within dynamic global power relations. The geopolitics of disaster coverage has an impact on everything from the representation of victims to the deployment of organizational resources, and it means that geographically and culturally "distant sufferers" are accorded much less attention than more proximate ones, and are therefore treated as unworthy victims. Such are its historical longevity and continuing geopolitical circumscription that this has become both institutionalized and naturalized in today's amoral "calculus of death," routinely deployed by news organizations. By contrast, the personalized and therefore worthier sufferers closer to home hail more cosmopolitan forms of citizenship among spectators who are able to act upon the suffering. Thus, despite the globalized and transnational nature of media and the disasters they cover, the nation state and its place within the geopolitical landscape of the world remain highly salient.

Our examination of the Burma cyclone and the China and Haiti earthquakes in Chapter 7, "Disaster Citizenship and the Assumption of State Responsibility," suggested that in all three disasters, the underlying story was about the role of the state and the international community in responding to these devastating disasters and aiding the victims: The coverage was premised on the idea that the state is responsible for coming to the aid of its beleaguered citizens and that any shortcoming in this respect amounted to a fundamental dereliction of duty. Citizens of Western democracies were, through the metadiscourses of democracy, embedded within the sphere of consensus, reminded of their own privilege, but also of the plight of victimized citizens whose lives, already blighted by failures of governance, were touched by disaster. Thus, while disaster coverage can build empathy, the cosmopolitan imagination is constructed through and bounded by the nationally and geopolitically inflected narratives of disasters. What emerges is a less-than-straightforward picture of the forms of disaster citizenship and cosmopolitanism made possible by media coverage, but one which allows for hopeful readings of the political possibilities of both persistent and emerging practices.

Second and related, disaster coverage articulates tensions between the individual and the collective, between state and citizens, and between the macropolitical and the micropolitical. On the one hand, in a globalized world, it is nonetheless the case that state actors continue to profoundly define and shape the available discourses about disasters, whether through the assumption of state responsibility, the selective staging of disaster relief efforts and the implicit metadiscourses about modes of governance that we have seen played out across a range of disasters. Disasters, then, frequently play out larger macropolitical dramas, highlighting, for example, the injustices of the military dictatorship in Burma, the changing place of China, Pakistan and the United States in the world and the complicity of the world's wealthy countries in the poverty and devastation of Haiti. These macropolitical dramas do not merely construct (and seek to fix) the meaning of *particular* nations, their histories and their trajectories, but they are also *relational* insofar as they characterize global power relations, making media institutions active participants in the maintenance and contestation of hegemony, though with the active involvement of political elites.

So it is certainly the case that elites continue to serve as the primary definers in news stories about disasters, and yet we also argued that disasters can be seen as providing an opportunity for the empowerment of individuals as citizens. This is because disasters give voice to the citizens of disaster-struck nations and regions and, in doing so, open up possibilities for raising questions of political import. We analyzed this in detail in Chapter 8, "Anger and Accountability in Disaster News," where we examined the coverage of the political potential of disasters from the bottom up. We focused on how the expression of anger in disaster coverage may open up a space for political contestation and critique through an emotionally charged language justified by the tragedy and loss experienced by the victims. As such, disaster coverage opens up a rare space for "ordinary people" to communicate injustice and direct criticism at power holders in society—corporations, governments and other social institutions that they otherwise rarely have the opportunity to directly hold to account. More profoundly, it allows for a democratized systemic critique of complex social processes.

A third key tension relates to the relationship between conventional journalistic ideals of objectivity and impartiality and an increasing attention to emotional expression and its capacity to generate compassion in audiences. This tension should be understood in the context of an affective turn in journalism studies and beyond, which has enabled a reconsideration of the conventional valorization of rationality and conceptions of journalistic professionalism premised on the reporter as a distanced, impartial and dispassionate observer. As we argued, journalism can be better made sense of when approached through a lens sensitized to the power and performance of emotion and which is, we also contend, now increasingly central to journalistic practices in reporting disasters as well as wider modes of cultural receptivity. Journalists themselves do heavy emotional labor in their professional work, but they also infuse their narratives with emotions. At the same time, discursive registers in the reporting of disaster are changing and challenging conventional understandings of journalistic objectivity, pointing to the ways in which journalistic witnessing and expression of emotions can help to stir audiences to action. The affective turn has been particularly important for theorizing disaster, insofar as the representation of distant suffering through journalistic witnessing and the attendant injunction to care are central to emerging normative understandings of the media as effective moral and emotional educators. Chapter 5, "Producing News, Witnessing Disasters," examined journalistic witnessing in more detail. We argued that such witnessing is informed by a seemingly entrenched "calculus of death." Through established newsroom routines and Western-centric cultural outlooks, journalists make institutionalized judgments about which disasters are worthy of inclusion on the news agenda, but there are also significant variations to time-honored "disaster scripts" on the basis of the specificities of each disaster, as exemplified in correspondents' reflections on the dramatic potential of earthquakes, which contain the potential for dramatic visual coverage and the drawn-out drama of aftershocks. At the same time, the staging and witnessing of disasters are also shaped by the intervention of political authorities who seek to publicly manage the visibility of both the tragedy and its relief. Journalists are compelled to negotiate these complexities while seeking to balance the tension between professional tenets of objectivity and impartiality and an evident, if less professionally explicit, injunction to care.

Chapter 4, "Emotional Discourses in Disaster News," drew on examples ranging from the recent Haiti earthquake to the 1999 Paddington rail crash to demonstrate that emotional discourses in disaster narratives fuel and shape the emotions of the local and global audiences. Chapter 8 expanded on this idea by revealing systematic patterns, born out of the conventions of journalism, which structure who is authorized to express anger and under what circumstances and constraints. We proposed that the subject position of the victim or the individual otherwise affected by the disaster allows for the articulation of anger—either through letters to the editor or in stories where ordinary people were used as sources. As a result of the shifting emotional style and forms of news discourse, there are new opportunities for the expression and the enactment of politicized emotion which, in turn, make possible new forms of accountability.

A final key tension that registers throughout this book and which comes to the fore in Chapter 9, "Transformations in Disaster Visibility," expresses the contemporary ferment in the field of media and communication scholarship more widely. This concerns the fast-changing relationship between traditional mainstream media and emerging forms of new media and how or to what extent this is opening up new forms of participatory communication and expanding the range of views and voices found across today's complex media ecology—including in times of disaster. Rather than suggesting that one side or the other is winning out in this encounter between top-down, mainstream, industrially organized media systems and bottom-up, grass-roots, culturally diverse and geospatially dispersed communication networks, we have sought to explore the encounter and interpenetrations of both in and through particular disasters. Together, both old and new media are, to borrow John Thompson's prescient phrase, now transforming the visibility of disasters around the world. This encounter is also helping to reconfigure relations of communication power in disasters and the forms and flows of media representation are radically complexifying, though they often remain no less hierarchically powered or politically contested. We have explored, for example, how news producers have sought to incorporate social media and UGC into their disaster packages and reports, and we examined the rapid rise and involvement of mobile telephony and social media in recent disasters. New technologies have undoubtedly enabled broader participation in the dissemination of news and images around disasters, while new forms of communication surveillance are also now entering into the formations and flows of the mainstream media, sometimes with visually and politically charged effects. Ultimately, the transformation of disaster visibility helps to focus the world's attention on human rights abuses, including the abuse of structurally determined humanitarian disasters in a world of plenty, and contributing wider endorsement to the institutions and normative culture of today's global "human rights regime," thereby supporting these same political opportunity structures.

Though these opportunity structures infused inside the media's visualization of disasters are profoundly shaped by enduring conventions of journalistic work and limited in a more overarching sense by prevailing power relations, there are growing cracks and fissures in the landscape of global mediated politics, as articulated in and through disaster coverage. In particular, we have shown that although disasters are shaped by global geopolitics, they can also increasingly give voice to disaster victims and citizen eyewitnesses, through conventional and new media forms, from conventional vox pop interviews to the increasing presence of new social media. As Chapter 4 documented, audience participation in the production of disaster news contributed to the formation of feeling communities in and through the media in the aftermath of the Victoria bushfire emergency in 2009. It highlighted the role of "ordinary people" as increasingly important news characters, sources and participants in disaster reporting, through the sharing of eyewitness news and the expression of the emotions of those affected by disasters. Transformations of visibility in disaster coverage, then, open up new opportunities for both participation and accountability through the complex interplay of "new" and "old" media.

If we can understand disasters in relation to these tensions, it is not merely a conceptual exercise but one which has profound social, political and economic consequences. Despite the certainty that disasters themselves are on the increase, the emerging developments explored across this book all contain notes of hopefulness in calling attention to possibilities for addressing disasters as global social, political, cultural and economic events in which, as the media coverage reminds us, we all have a stake. Though we are all rooted in particular bodies, locations, communities and identities, disaster coverage invites us to imagine, and develop compassion with, individuals beyond our own horizons of lived experience. The difficult, sometimes disconcerting and largely unrecognized emotional labor of journalists calls upon us to see the world in new ways, to enlarge both what we know and what we feel.

References

Abbott, C., Rogers, P., & Sloboda, J. (2006). *Global responses to global threats: Sustainable security for the 21st century.* Oxford, UK: Oxford Research Group.

Abu-Lughod, L., & Lutz, C. (2009). Emotion, discourse, and the politics of everyday life. In J. Harding & E. D. Pribram (Eds.), *Emotions: A cultural studies reader* (pp. 100–112). London, UK: Routledge.

Ahmed, N. (2010). *A user's guide to the crisis of civilization.* London: Pluto Press.

Ahmed, S. (2004). *The cultural politics of emotion.* Edinburgh: Edinburgh University Press.

Alexander, J. (1988). Culture and political crisis: Watergate and Durkheimian sociology. In J. Alexander (Ed.), *Durkheimian sociology: Cultural studies* (pp. 187–224). New York, NY: Cambridge University Press.

Alexander, J. (2006a). *The civil sphere.* Oxford, UK: Oxford University Press.

Alexander, J. (2006b). Cultural pragmatics: Social performance between ritual and strategy. In J. Alexander, B. Giesen, & J. Mast (Eds.), *Social performance: Symbolic action, cultural pragmatics, and ritual* (pp. 29–90). Cambridge, UK: Cambridge University Press.

Alexander, J. (2009). The democratic struggle for power: The 2008 presidential campaign in the USA. *Journal of Power, 2*(1), 65–88.

Alexander, J., Giesen, B., & Mast, J. (Eds.). (2006). *Social performance: Symbolic action, cultural pragmatics, and ritual.* Cambridge, UK: Cambridge University Press.

Alexander, J., & Jacobs, R. (1998). Mass communication, ritual and civil society. In T. Liebes & J. Curran (Eds.), *Media, ritual and identity* (pp. 23–41). London, UK: Routledge.

Alexander, J., & Smith, P. (2003). The strong program in cultural sociology: Elements of a structural hermeneutics. In J. Alexander (Ed.), *The meanings of social life: A cultural sociology* (pp. 11–26). Oxford, UK: Oxford University Press.

Allan, S. (2006). *Online news.* Maidenhead, UK: Open University Press.

Allan, S., & Thorsen, E. (Eds.). (2009). *Citizen journalism: Global perspectives.* London, UK: Peter Lang.

Aminzade, R., & McAdam, D. (2001). Emotions and contentious politics. In R. Aminzade, J. A. Goldstone, D. McAdam, E. J. Perry, W. H. Sewell Jr., S. Tarrow, & C. Tilly (Eds.), *Silence and voice in the study of contentious* politics (pp. 14–50). New York, NY: Cambridge University Press.

Amnesty International. (2009). *The state of the world's human rights: Amnesty International report 2009.* London, UK: Author.

Anderson, B. (1991). *Imagined communities: Reflections on the origin and spread of nationalism.* London, UK: Verso.

Archibugi, D. (1998). Principles of cosmopolitan democracy. In D. Archibugi, D. Held, & M. Köhler (Eds.), *Re-imagining political community* (pp. 198–230). Cambridge, UK: Polity Press.

Arendt, H. (1968). *Men in dark times.* New York, NY: Harcourt Brace Jovanovich.

Ashuri, T., & Pinchevski, A. (2009). Witnessing as a field. In P. Frosh & A. Pinchevski (Eds.), *Media witnessing: Testimony in the age of mass communication* (pp. 133–157). Basingstoke, UK: Palgrave Macmillan.

Bacon, W., & Nash, C. (2002). *News/worthy: How the Australian media cover humanitarian, aid and development issues.* Canberra, Australia: AusAid.

Bankoff, G. (2001). Rendering the world unsafe: Vulnerability as western discourse. *Disasters, 25*(1), 19–35.

Bantz, C. R., McCorkle, S., & Baade, R. C. (1980). The news factory. *Communication Research, 7*(1), 45–68.

Barbalet, J. (1998). *Emotion, social theory and social structure: A macro-sociological approach.* New York, NY: Cambridge University Press.

Barbalet, J. (2002). *Emotions and sociology.* Oxford, UK: Blackwell.

Barton, M. (1998). Journalistic gore: Disaster reporting and emotional discourse in the *New York Times*, 1852–1956. In P. N. Stearns & J. Lewis (Eds.), *An emotional history of the United States* (pp. 155–172). New York: New York University Press.

Baudrillard, J. (1988). *The ecstasy of communication.* New York, NY: Semiotext(e).

Bauman, Z. (2007). *Liquid times.* Cambridge, UK: Polity Press.

Beam, R., & Spratt, M. (2009). Managing vulnerability: Job satisfaction, morale and journalists' reactions to violence and trauma. *Journalism Practice, 3*(4), 421–438.

Beck, U. (1992). *Risk society.* London, UK: Sage.

Beck, U. (2000). *World risk society.* Cambridge, UK: Polity Press.

Beck, U. (2002). The silence of words and political dynamics in the world risk society. Retrieved from http://logosonline.home.igc.org/beck.htm

Beck, U. (2006). *Cosmopolitan vision.* Cambridge, UK: Polity Press.

Beck, U. (2009). *World at risk.* Cambridge, UK: Polity Press.

Beck, U., & Sznaider, N. (2006). Unpacking cosmopolitanism for the social sciences: A research agenda. *The British Journal of Sociology, 57*(1), 1–23.

Benhabib, S. (2004). *The rights of others: Aliens, residents, and citizens.* Cambridge, UK: Cambridge University Press.

Bennett, L. (1990). Towards a theory of press-state relations in the United States. *Journal of Communication, 40*(2), 103–127.

Bennett, L., Lawrence, R., & Livingston, S. (2007). *When the press fails: Political power and the news media from Iraq to Katrina.* Chicago: The University of Chicago Press.

Bennett, R., & Daniel, M. (2002). Media reporting of third world disasters: The journalist's perspective. *Disaster Prevention and Management, 11*(1), 33–42.

Ben-Porath, E. N., & Shaker, L. K. (2010). News images, race, and attribution in the wake of Hurricane Katrina. *Journal of Communication, 60*(3), 466–490.

Benthall, J. (1993). *Disasters, relief and the media.* London, UK: I. B. Tauris.

Berglez, P. (2008). What is global journalism? Theoretical and empirical conceptualisations. *Journalism Studies, 9*(6), 845–858.

Berglez, P., & Olausson, U. (2011). Intentional and unintentional transnationalism: Two political identities repressed by national identity in the news media. *National Identities, 13*(1), 35–49.

Bergmann, J., Egner, H., & Wulf, V. (2009). *Proposal for the establishment of a ZIF research group communicating disaster.* Bielefeld: University of Bielefeld.

Berkowitz, D. (Ed.) (1997). *Social meaning of news.* London, UK: Sage.

Berlant, L. (2004). Introduction: Compassion (and withholding). In L. Berlant (Ed.), *Compassion: The culture and politics of an emotion* (pp. 1–13). New York: Routledge.

Berrington, E., & Jemphrey, A. (2003). Pressures on the press: Reflections on reporting tragedy. *Journalism, 4*(2), 225–248.

Bianchi, A. (1999). Immunity versus human rights: The Pinochet case. *European Journal of International Law, 10,* 237–277.

Billig, M. (1995). *Banal nationalism.* London, UK: Sage.

Birkland, T. (1997). *After disaster: Agenda setting, public policy and focusing events.* Washington, DC: Georgetown University Press.

Bleiker, R., & Hutchison, E. (2008). Fear no more: Emotions and world politics. *Review of International Studies, 34,* 115–135.

Boin, A. (2005). From crisis to disaster: An integrative perspective. In R. Perry & E. Quarantelli (Eds.), *What is a disaster? New answers to old questions* (pp. 153–172). Philadelphia, PA: Xlibris.

Boin, A., & 't Hart, P. (2007). The crisis approach. In H. Rodriguez, H., E. Quarantelli, & R. Dynes (Eds.), *Handbook of disaster research* (pp. 42–54). New York, NY: Springer.

Boin, A., 't Hart, P., Stern, E., & Sundelius, B. (2005). *The politics of crisis management: Public leadership under pressure.* Cambridge, UK: Cambridge University Press.

Boltanski, L. (1999). *Distant suffering: Morality, media and politics.* UK: Cambridge University Press.

Boltanski, L. (2000). The legitimacy of humanitarian actions and their media representation: The case of France. *Ethical Perspectives, 7*(1), 3–16.

Bowman, S., & Willis, C. (2003). *We media: How audiences are shaping the future of news and information.* Reston, VA: The Media Center at the American Press Institute. Retrieved from http://www.hypergene.net/wemedia/download/we_media.pdf

Boyd-Barrett, O. (2005). A different scale of difference. *Global Media and Communication, 1*(1), 15–19.

Brown, B. (2009, June 6). Yes, I was callous. . . . Remarkable confession of guilt and shame by war reporter Ben Brown. Retrieved from http://www.dailymail.co.uk/news/article-1191231/Yes-I-callous—Remarkable-confession-war-reporter-Ben-Brown.html

Bulkley, K. (2010, June 18). Mobile technology takes centre stage in disaster relief. Retrieved from http://www.guardian.co.uk/activate/mobile-technology-disaster-relief

Burkitt, I. (2009). Powerful emotions: Power, government and opposition in the 'war on terror.' In J. Harding & E. D. Pribram (Eds.), *Emotions: A cultural studies reader* (pp. 157–169). London, UK: Routledge.

Butler, J. (2004). *Precarious life: The powers of mourning and violence.* London, UK: Verso.

Calhoun, C. (2008). The imperative to reduce suffering: Charity, progress, and emergencies in the field of humanitarian action. In M. Barnett & T. Weiss (Eds.), *Humanitarianism in question: Politics, power, ethics* (pp. 73–97). Ithaca, NY: Cornell University Press.

Campbell, D. (2004). Horrific blindness: Images of death in contemporary media. *Journal for Cultural Research, 8*(1), 55–74.

Carey, J. (1989). *Communication as culture: Essays on media and society.* Winchester, MA: Unwin Hyman.

CARMA. (2006). *The CARMA report on western media coverage of humanitarian disasters.* Retrieved from http://www.carma.com/research/#research

Carroll, N. (1997). Art, narrative, and emotion. In M. Hjort & S. Laver (Eds.), *Emotion and the arts* (pp. 190–211). New York, NY: Oxford University Press.

Castells, M. (2008). The new public sphere: Global civil society, communication networks and global governance. *Annals of the American Academy of Political Social Science, 616(1),* 78–93.

Castells, M. (2009). *Communication power.* UK: Oxford University Press.

Chandler, D. (2003). New rights for old? Cosmopolitan citizenship and the critique of state sovereignty. *Political Studies, 51,* 332–349.

Chouliaraki, L. (2006). *The spectatorship of suffering.* London, UK: Sage.

Chouliaraki, L. (2008a). The media as moral education: Mediation and action. *Media, Culture & Society, 30*(6), 831–852.

Chouliaraki, L. (2008b). The symbolic power of transnational media: Managing the visibility of suffering. *Global Media and Communication, 4*(3), 329–351.

Chouliaraki, L. (2009). Journalism and the visual politics of war and conflict. In S. Allan (Eds.), *Routledge companion to news and journalism* (pp. 520–533). London, UK: Routledge.

Chouliaraki, L. (2010). Witnessing in post-television news: Towards a new moral imagination. *Critical Discourse Studies, 7*(4), 305–319.

Clarke, C. (2005). Lessons from Katrina. *Poynter.org*, December 19. Retrieved from http://www.poynter.org/archived/covering-hurricanes/72800/lessons-from-katrina/

Clausen, L. (2004). Localizing the global: Domestication processes in international news production. *Media, Culture & Society, 26*(1), 25–44.

Cleaver, H. (2011). Japan's nuclear disaster deals a blow to Europe's atomic energy industry. *GlobalPost,* March 11. Retrived from http://www.globalpost.com/dispatch/news/regions/europe/germany/110315/japan-nuclear-germany-atomic-power

Cohen, A., Levy, M., Roeh, I., & Gurevitch, M. (1996). *Global newsroom, local audiences: A study of the Eurovision News Exchange.* London, UK: John Libbey.

Cohen, S. (2001). *States of denial: Knowing about atrocities and suffering.* Cambridge, UK: Polity Press.

Connell, I. (1998). Mistaken identities: Tabloid and broadsheet news discourse. *Javnost—The Public, 5*(3), 11–31.

Constable, M. (2008). Disaster mythology: Looting in New Orleans. *Disaster Prevention and Management, 17*(4), 519–525.

Cooper, G. (2011). *From their own correspondent? New media and the changes in disaster coverage: Lessons to be learnt.* Oxford, UK: University of Oxford, Reuters Institute for the Study of Journalism.

Corpus Ong, J. (2009). The cosmopolitan continuum: Locating cosmopolitanism in media and cultural studies. *Media, Culture & Society, 31*(3), 449–466.

Costera Meijer, I. (2001). The public quality of popular journalism: Developing a normative framework. *Journalism Studies, 2*(2), 189–205.

Cottle, S. (1993). *TV news, urban conflict and the inner city. Leicester,* UK: Leicester University Press.

Cottle, S. (2006a). *Mediatized conflict: Developments in media and conflict studies.* Maidenhead, UK: Open University Press.

Cottle, S. (2006b). Mediatized rituals: Beyond manufacturing consent. *Media, Culture & Society, 28*(3), 411–432.

Cottle, S. (2008). Journalism and globalization. In K. Wahl-Jorgensen & T. Hanitzsch (Eds.), *Handbook of journalism studies* (pp. 341–356). London, UK: Routledge.

Cottle, S. (2009a). *Global crisis reporting: Journalism in the global age.* Maidenhead, UK: Open University Press.

Cottle, S. (2009b). Global crises in the news: Staging new wars, disasters and climate change. *International Journal of Communication, 3,* 494–516.

Cottle, S. (2011a). Taking global crises in the news seriously: Notes from the dark side of globalization. *Global Media and Communication, 7*(2), 77–95.

Cottle, S. (2011b). Cell phones, camels and the global call for democracy. In J. Mair & R. Keeble (Eds.), *Mirage in the desert? Reporting the Arab Spring.* (pp. 196–210). Bury St Edmunds, Suffolk, UK: Arima.

Cottle, S. (2012). Mediatized disasters in the global age: On the ritualization of catastrophe. In J. Alexander, R. Jacobs, & P. Smith (Eds.), *Oxford handbook of cultural sociology* (pp. 259–283). Oxford, UK: Oxford University Press.

Cottle, S., & Ashton, M. (1999). From BBC newsroom to BBC newscentre: On changing technology and journalist practices. *Convergence: The Journal of Research into New Media Technologies, 5*(3), 22–43.

Cottle, S., & Lester, L. (Eds.). (2011). *Transnational protests and the media.* New York, NY: Peter Lang.

Cottle, S., & Nolan, D. (2007). Global humanitarianism and the changing aid-media field: 'Everyone was dying for footage.' *Journalism Studies, 8*(6), 862–878.

Cottle, S., & Rai, M. (2006). Between display and deliberation: Analyzing TV news as communicative architecture. *Media, Culture & Society, 28*(2), 163–189.

Couldry, N. (2003). *Media rituals: A critical approach.* London, UK: Routledge.

Dahlgren, P., & Sparks, C. (Eds.). (1992). *Journalism and popular culture.* London, UK: Sage.

Dart Center for Journalism & Trauma. (2005). Covering the tsunami: A frontline discussion. Retrieved from http://dartcenter.org/content/covering-tsunami-0

Dayan, D., & Katz, E. (1985). Television ceremonial events. *Society, 22*(4), 60–66.

Dayan, D., & Katz, E. (1992). *Media events: The live broadcasting of history.* Cambridge, MA: Harvard University Press.

Delanty, G. (2001). Cosmopolitanism and violence. The limits of global civil society. *European Journal of Social Theory, 4*(1), 41–52.

Deuze, M., Bruns, A., & Neuberger, C. (2007). Preparing for an age of participatory news. *Journalism Practice, 1*(3), 322–338.

Dillon, M., & Reid, J. (2000). Global governance, liberal peace and complex emergency. *Alternatives, 25*(1), 117–143.

Dovey, J. (2000). *Freakshow: First person media and factual television.* London, UK: Pluto Press.

Duffield, M. (2001). *Global governance and the new wars.* London, UK: Zed Books.

Duffield, M. (2007). *Development, security and unending war.* Cambridge, UK: Polity Press.

Durham, F. (2008). Media ritual in catastrophic time: The populist turn in television coverage of Hurricane Katrina. *Journalism, 9*(1), 95–116.

Dynes, R. (2000). The dialogue between Voltaire and Rousseau on the Lisbon earthquake: The emergence of a social science view. *International Journal of Mass Emergencies and Disasters, 18*(1), 97–115.

Edelstein, D. (2011, July 3). Activate 2011: Mobile phones in the developing world—Video. Retrieved from http://www.guardian.co.uk/media/video/2011/jul/04/activate-mobile-phones-developing-world-video

Edwards, C. (2009, May 4). *Responding to disasters in the age of social media.* Retrieved from http://www.newstatesman.com/print/200905040002

Ekström, M. (2002). Projections of power: Framing news, public opinion, and U.S. foreign policy. *Journalism, 3*(3), 259–282.

Ellis, J. (2000). *Seeing things: Television in the age of uncertainty.* London, UK: I. B. Tauris.

Eltringham, M. (2011, March 28). The new frontline is inside the newsroom [Web log post]. Retrieved from http://www.bbc.co.uk/journalism/blog/2011/03/how-the-newsroom-handles-confl.shtml

Energy Watch Group. (2007). *Oil report.* Retrieved from http://www.energywatchgroup.org

Entman, R. M. (2004). *Projections of power: Framing news, public opinion and US foreign policy.* Chicago: University of ChicagoPress.

Epstein, E. J. (1973). *News from nowhere.* New York, NY: Vintage Press.

Ettema, J. (1990). Press rites and race relations: A study of mass mediated ritual. *Critical Studies in Mass Communication, 7*(4), 309–331.

Fine, R. (2007). *Cosmopolitanism.* London, UK: Routledge.

Fox, K. (2011, July 23). Africa's mobile economic revolution. Retrieved from http://www.guardian.co.uk/technology/2011/jul/24/mobile-phones-africa-microfinance-farming

Franklin, B. (1997). *Newszak and news media.* London, UK: Arnold.

Friedman, S. M. (2011). Three Mile Island, Chernobyl, and Fukushima: An analysis of traditional and new media coverage of nuclear accidents and radiation. *Bulletin of the Atomic Scientists, 67*(5), 55–65.

Frosh, P. (2006). Telling presences: Witnessing, mass media, and the imagined lives of strangers. *Critical Studies in Media Communication, 23*(4), 265–284.

Frosh, P., & Pinchevski, A. (2009). Crisis-readiness and media witnessing. *The Communication Review 12*(3), 295–304.

Frosh, P., & Pinchevski, A. (Eds.). (2011). *Media witnessing: Testimony in the age of mass communication.* Basingstoke, UK: Palgrave Macmillan.

Gaddy, G. D., & Tanjong, E. (1986). Earthquake coverage by the western press. *Journal of Communication, 36*(2), 105–112.

Galtung, J., & Ruge, M. (1965). The structure of foreign news: The presentation of the Congo, Cuba and Cyprus crises in four Norwegian newspapers. *Journal of Peace Research, 2*(1), 64–90.

Gans, H. (1980). *Deciding what's news.* London, UK: Constable.

Garfield, G. (2007). Hurricane Katrina: The making of unworthy disaster victims. *Journal of African American Studies, 10,* 55–74.

Giddens, A. (1990). *The consequences of modernity.* Cambridge, UK: Polity Press.

Gillmor, D. (2004, July). *We the media.* Retrieved from http://www.oreilly.com/catalog/wemedia/book/index.csp

Glenn, J., & Gordon, J. (2007). *State of the future.* New York, NY: World Federation of United Nations Associations.

Global Humanitarian Forum. (2009). *Human impact report: Climate change—The anatomy of a silent crisis.* Geneva, Switzerland: Author.

Golding, P., & Elliott, P. (1999). Making the news. In H. Tumber (Ed.), *News: A reader.* (pp. 112–120). Oxford, UK: Oxford University Press.

Goodwin, J., & Jasper, J. (2003). *Rethinking social movements: Structure, meaning, and emotion.* Lanham, MD: Rowman & Littlefield.

Goodwin, J., Jasper, J., & Polletta, F. (2001). *Passionate politics: Emotions and social movements.* Chicago: The University of Chicago Press.

Gordon, S. (1990). Social structural effects on emotions. In T. D. Kemper (Ed.), *Research agendas in the sociology of emotions.* Albany: State University of New York Press.

Gould, D. (2001). Rock the boat, don't rock the boat, baby: Ambivalence and the emergence of militant AIDS activism. In J. Goodwin, J. M. Jasper, & F. Polletta (Eds.), *Passionate politics: Emotions and social movements* (pp. 383–391). Chicago: The University of Chicago Press.

Gould, D. (2009). *Moving politics. Emotion and ACT UP's fight against AIDS.* Chicago: The University of Chicago Press.

Gowing, N. (2009). *Skyful of lies and black swans: The new tyranny of shifting information power in crises.* Oxford, UK: University of Oxford, Reuters Institute for the Study of Journalism.

Grabe, M. E., & Bucy, E. P. (2009). *Image bite politics: News and the visual framing of elections.* Oxford, UK: Oxford University Press.

Gripsrud, J. (1992). The aesthetics and politics of melodrama. In P. Dahlgren & C. Sparks (Eds.), *Journalism and popular culture* (pp. 84–95). London, UK: Sage.

Gurevitch M., Levy, M., & Roeh, I. (1991). The global newsroom: Convergences and diversities in the globalisation of television news. In P. Dahlgren & C. Sparks (Eds.), *Communications and citizenship: Journalism and the public sphere in the new media age* (pp. 195–214). London, UK: Routledge.

Habermas, J. (1989). *The structural transformation of the public sphere.* Cambridge, UK: Polity Press.

Hall, S. (1973). The determinations of news photographs. In S. Cohen & J. Young (Eds.), *The manufacture of news* (pp. 226–243). London, UK: Constable.

Hall, S., Critcher, C., Jefferson, T., Clarke, J., & Roberts, B. (1977). *Policing the crisis.* Basingstoke, UK: Palgrave Macmillan.

Hallin, D. (1986). *The uncensored war: The media and Vietnam.* Oxford, UK: Oxford University Press.

Hallin, D. (1994). *We keep America on top of the world.* London, UK: Routledge.

Hammock, J., & Charny, J. (1996). Emergency response as morality play: The media, the relief agencies and the need for capacity building. In R. Rotberg & T. Weiss (Eds.), *From massacres to genocide: The media, public policy, and humanitarian crises* (pp. 115–135). Washington, DC: The Brookings Institution.

Hanitzsch, T., Hanusch, F., Mellado, C., Anikina, M., Berganza, R., Cangoz, I., Coman, M., Hamada, B., Hernandez, M. E., Karadjov, C., Moreira, S. V., Mwesige, P. G., Plaisance, P. L., Reich, Z., Seethaler, J. Skewes, E., Vardiansyah Noor, D., Kee Wang Yuen, E. (2011). Mapping journalism cultures across nations: A comparative study of 18 countries. *Journalism Studies*, 12(3), 273–293.

Hanusch, F. (2008). Valuing those close to us: A study of German and Australian quality newspapers reporting of death in foreign news. *Journalism Studies, 9*(3), 341–356.

Hanusch, F. (2010, October). *Visualising the dead in disasters. An international comparison of newspaper coverage of the 2010 Haiti earthquake.* Paper presented at the European Communication Conference, Hamburg, Germany.

Harcup, T., & O'Neill, D. (2001). What is news? Galtung and Ruge revisited. *Journalism Studies, 2*(2), 261–268.

Harding, J., & Pribram, E. D. (Eds.). (2009). *Emotions: A cultural studies reader.* London, UK: Routledge.

Harrington, S. (2008). Popular news in the 21st century: Time for a new critical approach? *Journalism, 9*(3), 266–284.

Harrison, J. (2006). *News.* London: Routledge.

Harrison, P., & Palmer, R. (1986). *News out of Africa: Biafra to band aid.* London, UK: Shipman.

Hawkins, V. (2008). *Stealth conflicts.* Houndmills, UK: Palgrave.

Heijmans, T. (2005). Sorry Pakistan, jullie ramp heeft het niet [Sorry Pakistan, your disaster just doesn't have it]. *De Volkskrant,* October 21. Retrieved from: http://www.volkskrant.nl/vk/nl/2668/Buitenland/article/detail/667640/2005/10/21/Sorry-Pakistan-jullie-ramp-heeft-het-niet.dhtml

Held, D. (2004). *The global covenant.* Cambridge, UK: Polity Press.

Held, D., Kaldor, M., & Quah, D. (2010). *The hydra-headed crisis.* London: London School of Economics.

Hellman, M., & Riegert, K. (2010). Transnational news and crisis reporting. The Indian Ocean tsunami on CNN and Swedish TV4. In U. Kivukuru & L. Nordin (Eds.), *After the tsunami: Crisis communication in Finland and Sweden* (pp. 127–148). Göteborg: Nordicom.

Herman, E., & Chomsky, N. (1988). *Manufacturing consent: The political economy of the mass media.* New York, NY: Pantheon Books.

Hochschild, A. (1979). Emotion work, feeling rules, and social structure. *American Journal of Sociology, 85,* 551–575.

Höijer, B. (2004). The discourse of global compassion: The audience and media reporting of human suffering. *Media, Culture and Society, 26*(4), 513–531.

Holmes, J., & Niskala, M. (2007). *Reducing the humanitarian consequences of climate change.* Retrieved from International Federation of Red Cross and Red Crescent Societies website: http://www.ifrc.org/Docs/News/opinion07/07101001/index.asp

Holmes, M. (2004). The importance of being angry: Anger in political life. *European Journal of Social Theory, 7*(2), 123–132.

Huang, T. (2011, April 4). Coverage of Japanese citizens' 'stoic' response to tragedy both accurate, stereotypical. Retrieved from http://www.poynter.org/how-tos/newsgathering-storytelling/diversity-at-work/126006/japanese-citizens-stoic-response-to-tragedy-both-accurate-stereotypical/

Hughes, A. L., Palen, L., Sutton, J., Liu, S. B., & Vieweg, S. (2008, May). 'Site-Seeing' in disaster: An examination of on-line social convergence. In F. Fiedrich & B. Van de Walle (Eds.), *Proceedings of the 5th International ISCRAM Conference.* Retrieved from http://www.cs.colorado.edu/~palen/Papers/iscram08/OnlineConvergenceISCRAM08.pdf

Hume, D. (1757). *Of tragedy.* Retrieved from http://www.ourcivilisation.com/smartboard/shop/humed/tragedy.htm

Hume, D. (1826). *Concerning Moral Sentiment. Appendix I.* Retrieved from http://ebooks.adelaide.edu.au/h/hume/david/h92pm/appendix1.html.

Hunt, D. (1999). *O. J. Simpson: Fact and fictions.* Cambridge, UK: Cambridge University Press.

Ignatieff, M. (1998). *The warrior's honor: Ethnic war and the modern conscience.* London, UK: Chatto and Windus.

Illouz, E. (2007). *Cold intimacies: The making of emotional capitalism.* Cambridge, UK: Polity Press.

International Federation of Red Cross and Red Crescent Societies. (2005). *World disasters report 2005.* Satigny/Vernier, Switzerland: ATAR Roto Press.

International Institute of Strategic Studies. (2007). *Strategic survey 2007: The annual review of world affairs.* London, UK: Author.

International Panel on Climate Change. (2007). *Climate change 2007: The physical science basis. Summary for policy makers.* Retrieved from http://www.ipcc.ch/pdf/assessment-report/ar4/wg1/ar4-wg1-spm.pdf

Iyengar, S. (1991). *Is anyone responsible? How television frames political issues.* Chicago: The University of Chicago Press.

Jääsaari, J. (2009). Restoring consensus after the blame game: The reception of Finnish tsunami crisis management coverage. In U. Kivikuru & L. Nord (Eds.), *After the tsunami. Crisis communication in Finland and Sweden* (pp. 57–82). Göteborg, Sweden: Nordicom.

Jacobs, R. (2000). *Race, media and the crisis of civil society: From Watts to Rodney King.* Cambridge, UK: Cambridge University Press.

Jaggar, A. (2009). Love and knowledge: Emotion in feminist epistemology. In J. Harding & E. D. Pribram (Eds.), *Emotions: A cultural studies reader* (pp. 50–68). London, UK: Routledge.

Jasper, J. (1997). *The art of moral protest: Culture, biography, and creativity in social movements.* Chicago: The University of Chicago Press.

Jasper J. (1998). The emotions of protest: Affective and reactive emotions in and around social movements. *Sociological Forum, 13*(3), 397–424.

Joye, S. (2009). The hierarchy of global suffering: A critical discourse analysis of television news reporting on foreign natural disasters. *Journal of International Communication, 15*(2), 45–61.

Joye, S. (2010). News media and the (de)construction of risk: How Flemish newspapers select and cover international disasters. *Catalan Journal of Communication, 2*(2), 253–266.

Kaldor, M. (2006). *New and old wars: Organized violence in a global era.* Cambridge, UK: Polity Press.

Kaldor, M. (2007). *Human security.* Cambridge, UK: Polity Press.

Kant, I. (2005). *Perpetual peace.* New York, NY: Cosimo Classics. (Original work published 1795)

Katz, E., & Liebes, T. (2007). No more peace! How disasters, terror and war have upstaged media events. *International Journal of Communication, 1*(1), 157–166.

Kayser-Bril, N. (2008, September 13). Emotion, victims and modern journalism. Retrieved from http://windowonthemedia.com/2008/09/emotion-journalism/

Keen, D. (2008). *Complex emergencies.* Cambridge, UK: Polity Press.

Kempe, M. (2003). Noah's flood: The Genesis story and natural disasters in early modern times. *Environment and History, 9*(2), 151–171.

Kim, H., & Cameron, G. (2011). Emotions matter in crisis: The role of anger and sadness in the publics' response to crisis news framing and corporate crisis response. *Communication Research, 38*(6), 826–855.

Kitch, C. (2003). Mourning in America: Ritual, redemption, and recovery in news narrative after September 11. *Journalism Studies, 4*(2), 213–224.
Kitch, C. (2009). Tears and trauma in the news. In B. Zelizer (Ed.), *The changing faces of journalism: Tabloidization, technology and truthiness* (pp. 29–39). New York, NY: Routledge.
Kitch, C., & Hume, J. (2008). *Journalism in a culture of grief.* New York, NY: Routledge.
Kivikuru, U. (2006). Tsunami communication in Finland. *European Journal of Communication, 21*(4), 499–520.
Kivikuru, U., & Nord, L. (2009). *After the tsunami: Crisis communication in Finland and Sweden.* Göteborg, Sweden: Nordicom.
Klein, N. (2007). *The shock doctrine: The rise of disaster capitalism.* London, UK: Allen Lane.
Knobloch-Westerwick, S., & Taylor, L. (2008). The blame game: Elements of causal attribution and its impact on siding with agents in the news. *Communication Research, 35*(6), 723–744.
Kristeva, J. (1991). *Strangers to ourselves.* Hemel Hempstead, UK: Harvester Wheatsheaf.
Kyriakidou, M. (2008). Rethinking media events in the context of a global public sphere: Exploring the audience of global disasters in Greece. *Communications, 33,* 273–291.
Kyriakidou, M. (2009). Imagining ourselves beyond the nation? Exploring cosmopolitanism in relation to media coverage of distant suffering. *Studies in Ethnicity and Nationalism, 9*(3), 481–496.
Lang, A., Dhillon, K., & Dong, Q. (1995). The effects of emotional arousal and valence on television viewers' cognitive capacity and memory. *Journal of Broadcasting and Electronic Media, 39*(3), 313–327.
Langer, L. (2003). Tabloid television and news culture: Access and representation. In S. Cottle (Ed.), *News, public relations and power* (pp. 135–151). London, UK: Sage.
Lappé, F., Collins, J., & Rosset, P. (1988). *World hunger: Twelve myths.* New York, NY: Earthscan.
Lee, C., Chan, J., Pan, Z., & So, C. (2005). National prisms of a global media event. In J. Curran & M. Gurevitch (Eds.), *Mass media and society* (pp. 259–309). London, UK: Edward Arnold.
Leith, D. (2004). *Bearing witness: The lives of war correspondents and photojournalists.* Milsons Point, NSW, Australia: Random House.
Levy, D., & Sznaider, N. (2002). Memory unbound. The Holocaust and the formation of cosmopolitan memory. *European Journal of Social Theory, 5*(1), 87–106.
Lewis, J., Inthorn, S., & Wahl-Jorgensen, K. (2005). *Citizens or consumers? What the media tell us about political participation.* Buckingham, UK: Open University Press.
Liebes, T. (1998). Television's disaster marathons: A danger for democratic processes? In T. Liebes & J. Curran (Eds.), *Media, ritual and identity* (pp. 71–84). London: Routledge.
Liebes, T., Kampf, Z., & Blum-Kulka, S. (2008). Saddam on CBS and Arafat on IBA: Interviewing the enemy on television. *Political Communication, 25*(3), 311–329.
Linklater, A. (1998). Cosmopolitan citizenship. *Citizenship Studies, 2*(1), 23–41.
Linklater, A. (2007). Distant suffering and cosmopolitan obligations. *International Politics, 44,* 19–36.
Lomborg, B. (Ed.). (2009). *Global crises, global solutions.* Cambridge, UK: Cambridge University Press.
Loseke, D. (2009). Examining emotion as discourse: Emotion codes and presidential speeches justifying war. *The Sociological Quarterly, 50*(3), 497–524.
Loyn, D. (2003, February 20). Witnessing the truth. Retrieved from http://www.opendemocracy.net/media-journalismwar/article_993.jsp
Lule, J. (2001). *Daily news, eternal stories: The mythological role of journalism.* New York, NY: Guilford Press.

Lull, J. (2007). *Culture-on-demand: Communication in a crisis world.* London, UK: BFI.

Lupton, D. (1998). *The emotional self.* London, UK: Sage.

Lyall, K. (2005, March 1). *The emotional toll of disaster reporting.* Retrieved from Dart Center for Journalism & Trauma website: http://dartcenter.org/content/emotional-toll-disaster-reporting-0

Lyman, P. (2004). The domestication of anger: The use and abuse of anger in politics. *European Journal of Social Theory, 7*(2), 133–147.

MacDonald, M. (2000). Rethinking personalization in current affairs journalism. In C. Sparks & J. Tulloch (Eds.), *Tabloid tales: Global debates over media standards* (pp. 251–266). Oxford, UK: Rowman & Littlefield.

Marcus, G. E., Neuman, W. R., & MacKuen, M. (2000). *Affective intelligence and political judgment.* Chicago: The University of Chicago Press.

Marshall, T. H. (1950). *Citizenship and social class, and other essays.* Cambridge, UK: Cambridge University Press.

Matthews, J., & Cottle, S. (2011). Television news ecology in the United Kingdom: A study of communicative architecture, its production and meanings. *Television & New Media.* Published first online. Retrieved from http://tvn.sagepub.com/content/early/2011/04/13/1527476411403630.

Mayes, T. (2000). Submerging in 'therapy news.' *British Journalism Review, 11*(4), 30–36.

McLean, I. (1999, February 12). How close were you to your dead child? Retrieved from *Times Higher Education* website: http://www.timeshighereducation.co.uk/story.asp?storyCode=145016§ioncode=26

McLean, I. (2007). Aberfan: No end of a lesson. Retrieved from http://www.historyandpolicy.org/papers/policy-paper-52.html#related

McLean, I., & Johnes, M. (2000). *Aberfan: Government and disasters.* Cardiff, UK: Welsh Academic Press.

McNair, B. (2000). *Journalism & democracy.* London: Routledge.

McNair, B. (2006). *Cultural chaos: Journalism, news and power in a globalized world.* London, UK: Routledge.

McNair, B. (2007). *An introduction to political communication* (4th ed.). London, UK: Routledge.

Minnear, L., Scott, C., & Weiss, T. (1996). *The news media, civil wars and humanitarian action.* Boulder, CO: Lynne Rienner.

Mody, B. (2010). *The geopolitics of representation in foreign news: Explaining Darfur.* Lexington, KY: Lexington Books.

Moeller, S. (1999). *Compassion fatigue: How the media sell disease, famine, war and death.* London, UK: Routledge.

Molotch, H., & Lester, M. (1974). News as purposive behavior: On the strategic use of routine events, accidents, and scandals. *American Sociological Review, 39*(1), 101–112.

Morgan, D. (2002). Pain: The unrelieved condition of modernity. *European Journal of Social Theory, 5*(3), 307–322.

Morrison, D. (1994). Journalists and the social construction of war. *Contemporary Record, 8*(2), 305–320.

Morrison, D. E., & Tumber, H. (1988). *Journalists at war—The dynamics of news reporting during the Falklands conflict* London, UK: Sage.

Muldoon, P. (2008). The moral legitimacy of anger. *European Journal of Social Theory, 11*(3), 299–314.

Murteira, H. (2004). The Lisbon earthquake of 1755: The catastrophe and its European repercussions. Retrieved from http://lisbon-pre-1755-earthquake.org/the-lisbon-earthquake-of-1755-the-catastrophe-and-its-european-repercussions/

Nabi, R. L. (2003). Exploring the framing effects of emotion. *Communication Research, 30*(2), 224–247.

Nash, K. (2008), Global citizenship as show business: The cultural politics of Make Poverty History. *Media, Culture & Society, 30*(2), 167–181.

Natsios, A. (1997). *US foreign policy and the four horsemen of the Apocalypse: Humanitarian relief in complex emergencies.* Westport, CT: Praeger.

Nelson, A., Sigal, I., & Zambrano, D. (2011). *Media, information systems and communities: Lessons from Haiti.* Miami, FL: Knight Foundation.

Newman, N. (2009). *The rise of social media and its impact on mainstream journalism.* Oxford, UK: University of Oxford, Reuters Institute for the Study of Journalism.

Nguyen, V. T. (2009). Remembering war, dreaming peace: On cosmopolitanism, compassion, and literature. *The Japanese Journal of American Studies, 20,* 149–174.

Nord, L., & Strömbäck, J. (2009). When a natural disaster becomes a political crisis. A study of the 2004 tsunami and Swedish political communication. In U. Kivikuru & L. Nord (Eds.), *After the tsunami. Crisis communication in Finland and Sweden* (pp. 17–40). Göteborg, Sweden: Nordicom.

Norris, P. (1995). The restless searchlight: Network news framing of the post Cold-War world. *Political Communication, 12*(4), 357–370.

Nussbaum, M. (1996a). Compassion: The basic social emotion. *Social Philosophy and Policy, 13,* 27–58.

Nussbaum, M. (1996b). Patriotism and cosmopolitanism. In J. Cohen (Ed.), *For love of country. Debating the limits of patriotism* (pp. 3–17). Boston, MA: Beacon.

Nussbaum, M. (2001). *Upheavals of thought: The intelligence of emotions.* Cambridge, UK: Cambridge University Press.

Nussbaum, M. (2003). Compassion and terror. *Daedalus, 132*(1), 10–26.

Odén, T., Ghersetti, M., & Wallin, U. (2009). Independence—Then adaption. How Swedish journalists covered the tsunami catastrophe. In U. Kivikuru & L. Nord (Eds.), *After the tsunami: Crisis communication in Finland and Sweden* (pp. 189–215). Göteborg, Sweden: Nordicom.

OECD. (2004). *Large-Scale disasters: Lessons learned.* Paris, France: Author.

O'Neill, D., & Harcup, T. (2009). News values and selectivity. In K. Wahl-Jorgensen & T. Hanitzsch (Eds.), *Handbook of journalism studies* (pp. 161–174). London, UK: Routledge.

Ost, D. (2004). Politics as the mobilization of anger: Emotions in movements and in power. *European Journal of Social Theory, 7*(2), 229–244.

Ó Tuathail, G. O., & Dalby, S. (1998). Introduction: Rethinking geopolitics: Towards a critical geopolitics. In G. O. Tuathail & S. Dalby (Eds.), *Rethinking geopolitics* (pp. 1–15). London, UK: Routledge.

Oxfam International. (2007). *Climate alarm: Disasters increase as climate change bites* (Briefing Paper No. 108). Retrieved from: http://www.oxfam.org.uk/resources/policy/climate_change/downloads/bp108_weather_alert.pdf

Oxfam International. (2009a). *The right to survive: The humanitarian challenge for the twenty-first century.* Retrieved from: http://www.oxfam.org/sites/www.oxfam.org/files/right-to-survive-report.pdf

Oxfam International. (2009b). *A billion hungry people.* Oxfam (Briefing Paper No. 127). Retrieved from: http://www.oxfam.org/sites/www.oxfam.org/files/bp127-billion-hungry-people-0901.pdf

Pantti, M. (2010). The value of emotion: An examination of television journalists' notions on emotionality. *European Journal of Communication, 25*(2), 168–181.

Pantti, M., & Bakker, P. (2009). Misfortunes, sunsets and memories: Non-Professional images in Dutch news media. *International Journal of Cultural Studies, 12*(5), 471–489.

Pantti, M., & Sumiala, J. (2009). Till death do us join: Media, mourning rituals and the sacred centre of the society. *Media, Culture & Society, 31*(1), 119–135.

Pantti, M., & Wahl-Jorgensen, K. (2007). On the political possibilities of therapy news: Media responsibility and the limits of objectivity in disaster coverage. *Studies in Communication Review, 1,* 3–25.

Pantti, M., & Wahl-Jorgensen, K. (2011). Not an act of God: Anger in British disaster coverage. *Media, Culture & Society, 33*(1), 105–122.

Pantti, M., & Wieten, J. (2005). Mourning becomes the nation: Television coverage of the murder of Pim Fortuyn. *Journalism Studies, 6*(3), 301–313.

Pedelty, M. (1995). *War stories: The culture of foreign correspondents.* New York, NY: Routledge.

Perera, S. (2010). Torturous dialogues: Geographies of trauma and spaces of exception. *Continuum, 24*(1), 31–45.

Perry, R. (2007). What is a disaster? In H. Rodriguez, E. Quarantelli, & R. Dynes (Eds.), *Handbook of disaster research* (pp. 1–15). New York, NY: Springer.

Peters, J. (2001). Witnessing. *Media, Culture and Society, 23*(6), 707–724.

Peters, J. (2011). An afterword: Torchlight red on sweaty faces. In P. Frosh & A. Pinchevski (Eds.), *Media witnessing: Testimony in the age of mass communication* (pp. 42–48). Houndmills, UK: Palgrave Macmillan.

Pew Research Center for the People and the Press. (1995, October 31). *A content analysis: International news coverage fits public's ameri-centric mood.* Retrieved from http://people-press.org/1995/10/31/a-content-analysis-international-news-coverage-fits-publics-ameri-centric-mood/

Philo, G. (1993). From Buerk to band aid. In J. Eldridge (Ed.), *Getting the message: News, truth and power* (pp. 104–125). London, UK: Routledge.

Price, M., Morgan, L., & Klinkforth, K. (2009, November 9). *NGOs as newsmakers: A new series on the evolving news ecosystem.* Retrieved from http://www.niemanlab.org/2009/11/ngos-as-news-makers-a-new-series-on-the-evolving-news-ecosystem/

Pupavac, V. (2004). War on the couch: The emotionology of the new international security paradigm. *European Journal of Social Theory, 7*(2), 149–170.

Quarantelli, E. (1989). The social science study of disasters and mass communication. In L. Walters, L. Wilkins, & T. Walters (Eds.), *Bad tidings: Communication and catastrophe.* Hillsdale, NJ: Lawrence Erlbaum Associates.

Quarantelli, E., Lagadec, P., & Boin, A. (2007). A heuristic approach to future disasters and crises: New, old, and in-between types. In H. Rodríguez, H., E. Quarantelli, & R. Dynes (Eds.), *Handbook of disaster research* (pp. 16–41). New York, NY: Springer.

Rahkonen, J., & Ahva, L. (2005). Hiipivä tuho [Sneaking destruction]. Journalismikritiikin vuosikirja 2005 [Yearbook of Journalism Criticism 2005]. *Tiedotustutkimus, 1,* 8–18.

Rancière, J. (2004). Who is the subject of the rights of man? *The South Atlantic Quarterly*, 103(2/3), 297–310.

Reddy, W. (2001). *The navigation of feeling: A framework for the history of emotion.* Cambridge, UK: Cambridge University Press.

Rentschler, C. (2004). Witnessing: US citizenship and the vicarious experience of suffering. *Media, Culture & Society, 26*(2), 296–304.

Rentschler, C. (2009). From danger to trauma: Affective labor and the journalistic discourse on witnessing. In P. Frosh & A. Pinchevski (Eds.), *Media witnessing: Testimony in the age of mass communication* (pp. 158–181). London, UK: Palgrave.

Richards, B. (2007). *Emotional governance: Politics, media and terror.* Basingstoke, UK: Palgrave Macmillan.

Richardson, J. (2005). News values. In B. Franklin, M. Hamer, M. Hanna, M. Kinsey, & J. Richardson (Eds.), *Key concepts in journalism studies* (pp. 173–174). London, UK: Sage.

Robertson, A. (2008). Cosmopolitanization and real time tragedy: Television news coverage of the Asian tsunami. *New Global Studies, 2*(2), 1–25.

Robertson, A. (2010). *Mediated cosmopolitanism: The world of television news.* Cambridge, UK: Polity Press.

Robertson, L. (2001). Showing emotion. *American Journalism Review, 23*(9), 45–47. Retrieved from http://www.ajr.org/article.asp?id=2386

Robinson, S. (2009). A chronicle of chaos: Tracking the news story of Hurricane Katrina from *The Times-Picayune* to its website. *Journalism, 10*(4), 431–450.

Rodríguez, H., Quarantelli, H.E, & Dynes, R. (Eds.). (2007). *Handbook of disaster research.* New York, NY: Springer.

Rohr, C. (2003). Man and natural disaster in the late Middle Ages: The earthquake in Carinthia and northern Italy on 25 January 1348 and its perception. *Environment and History, 9*(2), 127–149.

Rojecki, A, (2009). Political culture and disaster response: The great floods of 1927 and 2005. *Media Culture & Society, 31*(6), 957–976.

Rosenstiel, T., & Kovach, B. (2005, October 2). Media anger management. *The Washington Post.* Retrieved from http://www.journalism.org/node/138.

Rotberg, R. I., & Weiss, T. (Eds.). (1996). *From massacres to genocide: The media, public policy and humanitarian crises.* Washington, DC: The Brookings Institute.

Salmi, H. (Ed.). (1996). *Lopun alku. Katastrofien historiaa ja nykypäivää* [The beginning of the end. History of catastrophes and the present]. Turku, Finland: University of Turku.

Sambrook, R. (2010). *Are foreign correspondents redundant? The changing face of international news.* Oxford, UK: University of Oxford, Reuters Institute for the Study of Journalism.

Santos, J. (2009). *Daring to feel: Violence, the news media, and their emotions.* Lanham, MD: Lexington Books.

Schlesinger, P. (1987). *Putting reality together.* London, UK: Methuen.

Schudson, M. (1995). *The power of news.* Cambridge, MA: Harvard University Press.

Schudson, M. (2001). The objectivity norm in American journalism. *Journalism, 2*(2), 149–170.

Schudson, M. (2002). What's unusual about covering politics as usual. In B. Zelizer & S. Allan (Eds.), *Journalism after September 11* (pp. 36–47). New York, NY: Routledge.

Schulz, W. F. (1982). News structure and people's awareness of political events. *Gazette, 30,* 139–153.

Seaton, J. (2005). *Carnage and the media: The making and breaking of news about violence.* London, UK: Penguin.

Seib, P. (2002). *The global journalist: News and conscience in a world of conflict.* London, UK: Rowman & Littlefield.

Seitz, J. (2008). *Global issues: An introduction.* Oxford, UK: Blackwell.

Semetko, H., & Valkenburg, P. (2000). Framing European politics: A content analysis of press and television news. *Journal of Communication, 50*(2), 93–109.

Shaw, M. (1996). *Civil society and media in global crises: Representing distant violence.* London, UK: Pinter.

Shaw, M. (2005). *The new western way of war: Risk-Transfer war and its crisis in Iraq*. Cambridge, UK: Polity Press.

Sigelman, K., & Walkosz, B. J. (1992). Letters to the editor as a public opinion thermometer: The Martin Luther King holiday vote in Arizona. *Social Science Quarterly, 73,* 938–946.

Silk, J. (2000). Caring at a distance: (Im)partiality, moral motivation and the ethics of representation—Introduction. *Ethics, Place & Environment, 3*(3), 303–309.

Silverstone, R. (2007). *Media and morality: On the rise of mediapolis.* Cambridge, UK: Polity Press.

Simon, A. F. (1997). Television news and international earthquake relief. *Journal of Communication, 47*(3), 82–93.

Skitka, L. J., Bauman, C. W., & Mullen, E. (2006). Political tolerance and coming to psychological closure following the September 11, 2001, terrorist attacks: An integrative approach. *Personality and Social Psychology Bulletin, 30,* 743–756.

Slote, M. (2007) *The ethics of care and empathy.* New York, NY: Routledge.

Small, D. A., Lerner, J. S., & Fischhoff, B. (2006). Emotion priming and attributions for terrorism: Americans' reactions in a national field experiment. *Political Psychology, 27*(2), 289–298.

Smith, A. (2002). *The theory of moral sentiments.* Cambridge, UK: Cambridge University Press. (Original work published 1759).

Smith, D. (2009, October 22). *Africa calling: Mobile phone usage sees record rise after huge investment.* Retrieved from http://www.guardian.co.uk/technology/2009/oct/22/africa-mobile-phones-usage-rise

Soderlund, W., Briggs, E., Hildeebrandt, K., & Sidahmed, A. (2008). *Humanitarian crises and intervention: Reassessing the impact of mass media.* Sterling, VA: Kumarian Press.

Solomon, R. (1997). In defense of sentimentality. In M. Hjort & S. Laver (Eds.), *Emotion and the arts* (pp. 225–245). New York, NY: Oxford University Press.

Sontag, S. (2003). *Regarding the pain of others.* New York, NY: Farrar, Straus & Giroux.

Sorenson, J. (1991). Mass media and discourse on famine in the horn of Africa. *Discourse & Society, 2*(2), 223–243.

Sparks, C., & Tulloch, J. (Eds.). (2000). *Tabloid tales.* Lanham, MD: Rowman & Littlefield.

Spelman, E. (1989). Anger and insubordination. In A. Garry & M. Pearsall (Eds.), *Women, knowledge, and reality: Explorations in feminist philosophy* (pp. 263–274). Boston, MA: Unwin Hyman.

Spelman, E. (1997). *Fruits of sorrow: Framing our attention to suffering.* Boston, MA: Beacon Press.

Stallings, R. (1995). *Promoting risk: Constructing the earthquake threat.* New York, NY: Aldine de Gruyter.

Stallings, R. (1998). Disaster and the theory of social order. In E. L. Quarantelli (Ed.), *What is a disaster: Perspectives on the question* (pp. 127–145). New York, NY: Routledge.

Stanyer, J., & Davidson, S. (2011). The global human rights regime and the internet: Non-democratic states and the hypervisibility of oppression. In S. Cottle & L. Lester (Eds.), *Transnational protests and the media* (pp. 268–284). New York, NY: Peter Lang.

Stearns, C., & Stearns, P. (1986). *Anger: The struggle for emotional control in America's history.* Chicago: The University of Chicago Press.

Stevenson, N. (1999). *The transformation of the media: Globalization, morality and ethics.* London, UK: Longman.

Stevenson, N. (2004). Cosmopolitanism, culture and televised social suffering. In C. Paterson & A. Sreberny (Eds.), *International news in the twenty-first century* (pp. 225–242). Eastleigh, UK: John Libby.

Szerszynski, B., & Urry, J. (2006). Visibility, mobility and the cosmopolitan: Inhabiting the world from afar. *British Journal of Sociology, 57*(1), 113–131.

Sznaider, N. (1998). The sociology of compassion: A study in the sociology of morals. *Cultural Values, 2*(1), 117–139.

Tajfel, H. (1981). *Human groups and social categories.* Cambridge, UK: Cambridge University Press.

Taylor, J. (1998). *Body horror: Photojournalism, catastrophe, and war.* New York, NY: Manchester University Press.

Telford, J., Cosgrove, J., & Houghton, R. (2006). *Joint evaluation of the international response to the Indian Ocean tsunami: Synthesis report.* Retrieved from http://www.alnap.org/initiatives/tec/synthesis.aspx

Tester, K. (2001). *Compassion, morality and the media.* Buckingham, UK: Open University Press.

Thomas, J. (2002). *Diana's mourning. A people's history.* Cardiff, UK: University of Wales Press.

Thomas, R. J. (2011). Media morality and compassion for 'faraway others.' *Journalism Practice, 5*(3), 287–302.

Thompson, J. (1995). *The media and modernity.* Cambridge, UK: Polity Press.

Thompson, J. (2000). *Political scandal: Power and visibility in the media age.* Cambridge, UK: Polity Press.

Thompson, J. (2006). The new visibility. *Theory, Culture & Society,* (6), 31–51.

Thompson, S. (2006). Anger and the struggle for justice. In S. Clarke, P. Hoggett, & S. Thompson (Eds.), *Emotions, politics and society* (pp. 123–144). London, UK: Routledge.

Thumim, N. (2010). Self-Representation in museums: Therapy or democracy? *Critical Discourse Studies, 7*(3), 291–304.

Tierney, K. (2007). From the margins to the mainstream? Disaster research at the crossroads. *Annual Review of Sociology, 33,* 503–525.

Tierney, K., Bevc, C., & Kuligowski, E. (2006). Metaphors matter: Disaster myths, media frames and their consequences in Hurricane Katrina. *The Annals of the American Academy, 604,* 57–81.

Tilly, C. (2010). The blame game. *The American Sociologist, 41,* 382–389.

Tuchman, G. (1972). Objectivity as strategic ritual: An examination of newsmen's notions of objectivity. *The American Journal of Sociology, 77,* 660–679.

Tuchman, G. (1978). *Making news: A study in the construction of reality.* New York, NY: Free Press.

Tumber, H. (2004). Prisoners of news values? Journalists, professionalism, and identification in times of war. In S. Allan & B. Zelizer (Eds.), *Reporting war* (pp. 190–206). London, UK: Routledge.

Tumber, H. (2006). The fear of living dangerously: Journalists who report on conflict. *International Relations, 20*(4), 439–451.

Tumber, H., & Prentoulis, M. (2003). Journalists under fire: Subcultures, objectivity and emotional literacy. In D. Thussu & D. Freedman (Eds.), *War and the media: Reporting conflict 24/7* (pp. 215–230). London, UK: Sage.

United Nations. (2009, June). *United Nations conference on the world financial and economic crisis and its impacts on development.* Retrieved from http://www.un.org/ga/president/63/interactive/financialcrisis/PreliminaryReport210509.pdf (accessed 7.12.11)

United Nations Development Program. (2009). *The threat posed by the economic crisis to universal access to HIV services for migrants.* Retrieved from http://www.unaids.org/en/knowledgecentre/resources/features/archive/2009/200908 (accessed 7.12.11)

United Nations Environmental Program. (2001). *What are natural hazards and how they can become natural disasters. Awareness and preparedness for emergencies on a local level.* Retrieved from http://www.unep.fr/scp/sp/disaster/natural.htm (accessed 7.12.11)

United Nations Environmental Program. (2007). *Vulnerable in a world of plenty. Global environment outlook 4* (Fact Sheet no. 14). Retrieved from http://www.unep.org/geo/ge04/ (accessed 7.12.11)

United Nations, Food and Agricultural Organization. (2009). *State of the world's forests.* New York, NY.

United Nations Foundation. (2011a). *Disaster relief 2.0: Information sharing in humanitarian emergencies.* Retrieved from http://www.unfoundation.org/assets/pdf/disaster-relief-20-report.pdf (accessed 7.12.11)

United Nations Foundation. (2011b). Fact sheet re: Disaster relief 2.0: Information sharing in humanitarian emergencies. Retrieved from http://www.unfoundation.org/assets/pdf/disaster-relief-fast-facts.pdf (accessed 7.12.11)

United Nations Inter-Agency Project on Human Trafficking. (2009, July 20). *Cambodia: Exodus to the sex trade? Effects of the global financial crisis.* Retrieved from http://www.no-trafficking.org/reports_docs/siren/siren_cb-04.pdf (accessed 7.12.11)

United Nations, Secretary-General. (2010, August 19). Secretary-General tells General Assembly Pakistan's disaster from devastating floods 'one of the greatest tests of global solidarity in our times. Retrieved from http://www.un.org/News/Press/docs/2010/sgsm13065.doc.htm (accessed 7.12.11)

Urry, J. (2003). *Global complexity.* Cambridge, UK: Polity Press.

van Dijk, T. A. (2009). News, discourse and ideology. In K. Wahl-Jorgensen & T. Hanitzsch (Eds.), *The handbook of journalism studies* (pp. 191–204). New York, NY: Routledge.

van Ginneken, J. (1998). *Understanding global news: A critical introduction.* London, UK: Sage.

van Zoonen, L. (2005). *Entertaining the citizen: When politics and popular culture converge.* Lanham, MD: Rowman & Littlefield.

Virilio, P. (2007). *The original accident.* Cambridge, UK: Polity Press.

Volkmer, I. (2002). Journalism and political crises in the global network society. In B. Zelizer & S. Allan (Eds.), *Journalism after September 11* (pp. 235–246). London, UK: Routledge.

Wahl-Jorgensen, K. (2007). *Journalists and the public: Newsroom culture, letters to the editor, and democracy.* Creskill, NJ: Hampton Press.

Wahl-Jorgensen, K. (2009). On the newsroom-centricity of journalism ethnography. In L. Bird (Ed.), *Journalism and anthropology* (pp. 21–35). Bloomington: Indiana University Press.

Walter, T. (2006). Disaster, modernity, and the media. In K. Garces-Foley (Ed.), *Death and religion in a changing world* (pp. 265–282). Armonk, NY: M. E. Sharpe.

Walter, T., Littlewood, J., & Pickering, M. (1995). Death in the news: The public invigilation of private emotion. *Sociology, 29*(4), 579–596.

Ward, S. (2010, October 14). *Emotion in reporting: Use and abuse.* Retrieved from http://ethics.journalism.wisc.edu/2010/10/14/ethics-center-co-authors-report-on-nonprofit-journalism-10/

Wardle, C. (2007). Monsters and angels: Visual press coverage of child murders in the US and UK, 1930–1990. *Journalism, 8*(4), 281–302.

Wardle, C., & Williams, A. (2008). *UGC @ the BBC: Understanding its impact upon contributors, non-contributors, and BBC News.* Cardiff, UK: Cardiff School of Journalism, Media and Cultural Studies.

Warner, M. (2002). *Publics and counterpublics.* New York, NY: Zone Books.

Weaver, D. (Ed.). (1997). *The global journalist: News people around the world.* Creskill, NJ: Hampton Press.

Wiesslitz, C., & Ashuri, T. (2011). 'Moral journalists': The emergence of new intermediaries of news in an age of digital media. *Journalism, 12*(8), 1035–1051.

Wilkinson, I. (2005). *Suffering: A sociological introduction.* Cambridge, UK: Polity Press.

Williams, A., Wahl-Jorgensen, K., & Wardle, C. (2011). 'More real and less packaged': Audience discourse on amateur news content and its effects on journalism practice. In K. Andén-Papadopoulos & M. Pantti (Eds.), *Amateur images and global news* (pp. 193–209). Bristol, UK: Intellect.

Willis, J. (2003). *The human journalist: Reporters, perspectives, and emotions.* Westport, CT: Praeger.

Wolfsfeld, G. (1997). *Media and political conflict.* Cambridge, UK: Cambridge University Press.

Woodward, K. (2004). Calculating compassion. In L. Berlant (Ed.), *Compassion: The culture and politics of an emotion.* New York, NY: Routledge.

World Disasters Report (2005). *World Disasters Report 2005: Focus on information in disasters.* Geneva, Switzerland: International Federation of Red Cross and Red Crescent Societies (IFRC).

World Health Organization. (2007). *A safer future: Global public health security.* Geneva, Switzerland: Author. Retrieved from http://www.who.int/whr/2007

Zelizer, B. (2007). On 'having been there': 'Eyewitnessing' as a journalistic key word. *Critical Studies in Media Communication, 24*(5), 408–428.

Žižek. S. (2009). *First as tragedy, then as farce.* London, UK: Verso.

Žižek, S. (2010). *Living in the end times.* London, UK: Verso.

Index

Simon Cottle, *General Editor*

From climate change to the war on terror, financial meltdowns to forced migrations, pandemics to world poverty, and humanitarian disasters to the denial of human rights, these and other crises represent the dark side of our globalized planet. They are endemic to the contemporary global world and so too are they highly dependent on the world's media.

Each of the specially commissioned books in the *Global Crises and the Media* series examines the media's role, representation, and responsibility in covering major global crises. They show how the media can enter into their constitution, enacting them on the public stage and thereby helping to shape their future trajectory around the world. Each book provides a sophisticated and empirically engaged understanding of the topic in order to invigorate the wider academic study and public debate about the most pressing and historically unprecedented global crises of our time.

For further information about the series and submitting manuscripts, please contact:

Dr. Simon Cottle
Cardiff School of Journalism
Cardiff University, Room 1.28
The Bute Building, King Edward VII Ave.
Cardiff CF10 3NB
United Kingdom
CottleS@cardiff.ac.uk

To order other books in this series, please contact our Customer Service Department at:

(800) 770-LANG (within the U.S.)
(212) 647-7706 (outside the U.S.)
(212) 647-7707 FAX

Or browse online by series at:

www.peterlang.com